DESIGNING GOTHAM

CONFLICTING WORLDS:
NEW DIMENSIONS OF THE AMERICAN CIVIL WAR
T. Michael Parrish, Series Editor

DESIGNING GOTHAM

WEST POINT ENGINEERS AND
THE RISE OF MODERN NEW YORK
1817–1898

JON SCOTT LOGEL

Louisiana State University Press
Baton Rouge

Published with the assistance of the V. Ray Cardozier Fund

Published by Louisiana State University Press
Copyright © 2016 by Louisiana State University Press
All rights reserved
Manufactured in the United States of America
First printing

DESIGNER: Michelle A. Neustrom
TYPEFACE: Adobe Caslon Pro
PRINTER AND BINDER: McNaughton & Gunn, Inc.

LIBRARY OF CONGRESS CATALOGING-IN-PUBLICATION DATA

Names: Logel, Jon Scott, 1968– author.
Title: Designing Gotham : West Point engineers and the rise of modern New York, 1817–1898 / Jon Scott Logel.
Description: Baton Rouge : Louisiana State University Press, [2016] | Series: Conflicting worlds : new dimensions of the American Civil War | Includes bibliographical references and index.
Identifiers: LCCN 2016005788| ISBN 978-0-8071-6372-6 (cloth : alkaline paper) | ISBN 978-0-8071-6373-3 (pdf) | ISBN 978-0-8071-6374-0 (epub) | ISBN 978-0-8071-6375-7 (mobi)
Subjects: LCSH: City planning—New York (State)—New York—History—19th century. | Civil engineering—New York (State)—New York—History—19th century. | Social change—New York (State)—New York—History—19th century. | Civil engineers—New York (State) —New York—Biography. | Military cadets—New York (State) —West Point—Biography. | United States Military Academy—Biography. | United States. Army—Biography. | United States—History—Civil War, 1861–1865—Biography. | New York (N.Y.)—History—1775–1865. | New York (N.Y.) —History—1865–1898.
Classification: LCC HT168.N5 L64 2016 | DDC 307.1/21609747—dc23 LC record available at https://lccn.loc.gov/2016005788

CONTENTS

6

TOWARD CONSOLIDATION
Bridges, Bosses, and Brooklyn | 123

7

REDEMPTION IN POSTBELLUM
GOTHAM | 150

8

THE EMERGENCE OF MODERN AMERICA
IN WEST POINT'S NEW YORK | 178

Photographs follow page 84.

CHRONOLOGY

1802   United States Military Academy (USMA) founded.

1815   Sylvanus Thayer and William McRee tour France.

1817   Thayer becomes superintendent of USMA.

1823   George S. Greene graduates from USMA.

1824   U.S. Congress passes the Survey Act of 1824.

Dennis Hart Mahan graduates from USMA.

1825   Erie Canal opened.

1831   Mahan begins tenure as professor of engineering at USMA.

1833   Thayer resigns as superintendent of USMA.

David Douglass appointed as chief engineer of the Croton Aqueduct.

William H. Sidell graduates from USMA.

1842   John Newton, Gustavus Woodson Smith, and Mansfield Lovell graduate from USMA.

First Croton Aqueduct completed.

1845   William F. "Baldy" Smith and Fitz John Porter graduate from USMA.

1846   George B. McClellan graduates from USMA.

Mexican-American War begins.

1847   Egbert L. Viele graduates from USMA.

1848   Mexican-American War ends.

1849   Croton Aqueduct Department created.

Alfred Craven appointed chief engineer and commissioner of Croton Aqueduct Department.

1850   Gouverneur Kemble Warren graduates from USMA.

1852   Henry Warner Slocum graduates from USMA.

American Society of Civil Engineers (ASCE) holds first meeting at the Croton Aqueduct office.

Army Corps of Engineers begins clearing Hell Gate in the East River.

1856   Egbert L. Viele appointed as engineer-in-chief of Central Park, drafts first plan for the park.

Alfred Craven appoints Greene as a Croton Aqueduct engineer.

1857   New York state legislature passes the Charter of 1857.

Frederick Law Olmsted appointed superintendent of Central Park.

1858   Olmsted and Calvert Vaux's Greensward Plan selected as the design for Central Park; Viele dismissed as engineer-in-chief.

1861   Start of the Civil War; Viele promoted from captain to brigadier general.

1862   McClellan relieved as commander of the Army of the Potomac.

1862–63   Brigadier Egbert L. Viele serves as military governor of Norfolk, Virginia.

General Fitz John Porter court-martialed for Union defeat at the Battle of Second Manassas.

1863   Generals George S. Greene, Gouverneur Kemble Warren, and Henry Warner Slocum fight for the Union Army at the Battle of Gettysburg.

New York City draft riots.

1864   McClellan loses to Lincoln in presidential election.

1865   Civil War ends.

Emily Warren and Washington Roebling married.

Viele publishes his 1864 "Water Map" of New York.

1866   Greene returns to New York and Croton Aqueduct Department.

Henry Warner Slocum becomes a Democrat and moves to Brooklyn.

James Laurie and others revive ASCE in New York City.

Upper West Side residents form the West End Association.

1867   ASCE resumes regular meetings in New York City.

1868   Greene replaces Craven as chief engineer of the Croton Aqueduct
        Department.

        Slocum elected as congressman for Brooklyn.

1869   Brooklyn Bridge construction begins.

1870   New York state legislature passes the "Tweed" charter of 1870.

        McClellan appointed as chief engineer of the Docks Department.

1871   Mahan takes his own life on a Hudson River steamboat.

1872   Emily Warren Roebling begins issuing her husband's directives for
        construction of the Brooklyn Bridge.

1873   New York state legislature passes the Charter of 1873.

        McClellan resigns as chief engineer of the Docks Department.

1875   Fitz John Porter appointed as commissioner of public works.

        William "Baldy" Smith appointed as commissioner of police.

1879   West End Association reorganized as the West Side Association and
        lobbies for elevated railways in the Upper West Side.

1880   Riverside Avenue completed.

1883   Brooklyn Bridge opened.

1884   Porter appointed as commissioner of police.

1885   General John Newton successfully removes the last obstacle,
        "Flood Rock," at Hell Gate in the East River.

1895   Harlem River Canal opened.

1897   Grant's Tomb completed on the Upper West Side.

1898   Greater New York City consolidated.

1902   USMA centennial celebration.

DESIGNING GOTHAM

INTRODUCTION

Nineteenth-Century New York and West Point

Walked tonight on the west side. The viaduct railroad, the Brooklyn Bridge, stone piers, and a river street 250 feet wide, the blowing up of Hellgate reefs—these changes all now under way, will make this a new city within ten years.

—GEORGE TEMPLETON STRONG,

5 June 1871, from *The Diary of George Templeton Strong,* Vol. 4

In June 1871, George Templeton Strong went for an evening stroll along the West Side of Manhattan. Later that night, he noted in his famous diary the tangible harbingers of progress in the growing metropolis. Elevated railroads for mass transportation, the Brooklyn Bridge, Riverside Drive, and the clearing of Hell Gate in the East River were all projects underway at the direction of engineers. Among those responsible for these "changes" were men educated in Professor Dennis Hart Mahan's program of engineering at the United States Military Academy.[1] John Newton, class of 1842, was the engineer who successfully blew up the "Hellgate reefs." Egbert L. Viele, class of 1847, first engineer-in-chief of Central Park and a West Side "booster," had proposed alternative designs for the "viaduct railroad" and the "250 feet wide" river street that became Riverside Drive.[2] Henry Warner Slocum, class of 1852, was a Brooklyn Democratic politician and congressman who sought to reform the corruption surrounding the Brooklyn Bridge construction. Newton, Viele, and Slocum were the most visible West Point graduates who helped to make New York "a new city" in the two decades following the Civil War. Whether they performed duties as engineers, elected representatives, or political appointees in the new metropolis, these former students of Mahan joined with other politicians, engineers, and urban boosters to create the Victorian New York that George Templeton Strong marveled at in June of 1871.

The West Pointers who helped build the city that Strong described were part of the transformation that made New York City the center of the United States in the nineteenth century. Between 1825 and 1898, New York evolved rapidly from a vital Atlantic port of trade into the epicenter of American commerce and culture. In 1825, New York was mainly old Manhattan with a population of about 165,000 concentrated below 13th Street at the tip of the island. At the century's end, John Randel's grid covered Manhattan from end to end, and more than 1.5 million people lived in New York (including the Bronx).[3] In the course of this commercial growth and cultural development, New Yorkers settled throughout and beyond Manhattan Island, making the city the epitome of a nineteenth-century metropolis. By 1852, New Yorkers believed that their city would grow to be the largest in the world, leading the population growth in the nation, "sounding from the Atlantic to the Pacific shores." Supporters boasted: "New York is a commercial city, a mart of the sea . . . for the merchants and merchandise of the world. On one side it is bounded by a narrow arm of the sea, and on the other by a broad and noble river." It was a place where "men [came] with what is new, and . . . to see what is new." New York made it possible for the masses "to spread and elicit knowledge."[4] As Herbert Croly argued in 1903, New York was "unquestionably becoming the most highly organized and the most distinguished collective expression of American social life."[5] Moreover, these commercial and social changes underpinned an aggressive progression in the rise of a uniquely American municipal government seeking control over the dynamic forces at work throughout the city. Nineteenth-century Americans reviled and revered New York in transformation: a city that encompassed both the tradition of national progress through westward expansion and economic growth on the one hand, and intense urbanization and centralization of authority and power on the other.[6]

New York's rise was nothing short of remarkable, especially when compared to its American peers—Boston, Philadelphia, and later Chicago. Since its founding as a Dutch trading post in the seventeenth century, New York City has weathered multiple crises, technological evolution, and significant demographic shifts, all demonstrating the city's resilience as America's leading metropolis. Three important factors contributed to New York's rise over the last three centuries. First, New York's deepwater port and the Hudson River created a superior line of communication to the rich North American interior. Second, New York's geographic advantages enabled the city to become a key transportation hub for manufacturing and product distribution, which fueled intense growth in commerce and trade in Manhattan. Lastly, the superior geographic position and economic good fortune attracted waves of aspiring workers and immigrants, especially over the course of the nineteenth century. With over 819,000 people living in the city in 1860, New York was the biggest city in the United States, home to 250,000 more inhabitants than the second-most populated city, Philadelphia. Economies of scale and its central location propelled nineteenth-century New York to become a global center for manufacturing garments, refining sugar, and publishing. Manufacturing employed some 43,340 New Yorkers by mid-century, again more than any other American city. Publishing houses quickly reproduced and printed the latest books arriving from London, written by the most popular English authors.[7] Possessing the means to produce and earn income, nineteenth-century New Yorkers were also a ready consumer base for domestic and imported products coming into the port. Manhattan became a cauldron of business, commerce, manufacturing, and finance, while, at the same time, an ascendency in municipal management, services, and infrastructure accompanied the city's growth. It was a city undergoing a metamorphosis at a scale and speed never before seen, ultimately transforming New York City into a uniquely American metropolitan powerhouse.

While much of this urban transformation has been documented, the role of individuals educated at the U.S. Military Academy on the "S" turn of the Hudson River remains mostly unexplored or unnoticed in New York history.[8] In addition to Newton, Viele, and Slocum, men such as George S. Greene and Fitz John Porter studied an engineering curriculum designed to produce army officers, served in the United States Army, and then, as civilians, came to New York City to advance their careers and status through the creation of Victorian Gotham. The national Military Academy at West Point trained future officers not only to lead soldiers and fight America's wars, but to design bridges and draw topographical maps. These Academy graduates, collectively known as the "Long Gray Line," brought water to the city, created Central Park, and represented the interests of New Yorkers in the U.S. Congress. Thus, the careers of these West Pointers culminated not in the combat of Mexico, Antietam, and Gettysburg, but in the construction of the Croton Aqueduct, the development of the Upper West Side, and the creation of metropolitan New York, all of which occurred fifty miles south of the national military institution.[9]

These examples of West Pointers in their civilian careers portend a deeper influence of West Point in nineteenth-century New York, and beyond that, the country as a whole. West Point produced a body of experts who advanced science, technology, and engineering throughout the nation, most notably through the Army Corps of Engineers.[10] Histories of the U.S. Military Academy and New York City make brief references to the involvement of West Point graduates in the construction of internal improvement projects, the professionalization of engineers, and the politics of New York.[11] West Point's influence on civil engineering became manifest through its graduates, who proved indispensable to the city's rise to prominence in the 1800s. Examining the civilian careers of West Point graduates in New York reveals that these alumni were more than just typical Academy-trained army officers who became successful in civilian life. Military historian William Skelton states that West Point furnished "scores" of graduates for civilian engineering and topographical careers.[12] Those graduates who ended up in New York City formed influential relationships with the political and social powers of Gotham. Together, they produced the engines of physical, social, and political change in the antebellum and postbellum city.

Two chief identities defined the subsequent lives of the nineteenth-century graduates of West Point: the first as military cadets, and the second as army officers. For the West Pointers who left the army and went to New York, they took on at least one more identity, that of New Yorkers. Merely graduating from West Point and serving in the army did not constitute success as a civilian in the city. However, the West Pointers who were successful in New York leveraged their education and military experiences to contribute to, and at times exploit, New York's social, economic, and political transformation. As part of nineteenth-century New York society, these former military men ambitiously sought to sway the course of the city's urban growth before and after the Civil War. This book explores the connections among the three central experiences of the West Pointers—as cadets, military officers, and New Yorkers—and the influence they had on the transformation of New York City between 1825 and 1898, from the opening of Erie Canal through the consolidation of Greater New York.

This is not to say that there was a cause and effect between the efforts of the West Point engineers and the development of the city, but rather, the relationship between men and place was an interacting evolution of thoughts, values, actions, and development. Recently, Carl Smith, the well-known scholar and historian of urban America, wrote that "a city is as much an *infrastructure of ideas* as it is a gathering of people, a layout of streets, . . . or a collection of political, economic, and social institutions." He asserts that this "infrastructure of ideas neither precedes nor follows the building of a physical and social infrastructure, but is inseparable from them."[13] This study reaffirms that thesis by exploring the simultaneous emergence of science and technology at West Point, the ascendance of American engineering expertise, the changing urban landscape during New York's rapid rise to prominence, and the ways those factors combined at the dawn of modern America between the end of the Civil War and World War I. During this half-century, the United States underwent critical changes in how it conceived of itself as a nation and its place in the world. In New York, problems stemming from exponential population growth, increased demand for food production, and the associated "filth" that resulted from these increases all had to be solved. Modern transformations included changing laws and municipal codes to manage new waterworks, sewers, and roads as well as provide sanitation in the burgeoning cities. Additionally, city leaders had to create housing regulations, provide public education, and address health care in the face of disease and epidemics. During a period marked by disorder and chaos, New Yorkers created processes, institutions, and physical structures to impose order in their emergent modern world. American engineering and the rise of American cities were just one facet—but a vital one—of this transformative period.[14] More than merely an examination of the connection between ideas and physical structures, this book is primarily a study of how human relationships forged uniquely American ideas in the development of engineering, professionalism, and civic administration.

NEW YORK'S AND WEST POINT'S RISE IN THE ANTEBELLUM ERA

The completion of the Erie Canal in 1825 poised New York City for its rise as the center of American commerce and, indeed, nineteenth-century American culture. Just a day's steamboat ride up the Hudson lay the foundation of civil engineering in the antebellum United States.[15] Between 1802 and 1828, West Point was the only institution in America teaching civil engineering. Graduates who did not desire a career in the military could resign their commissions and make a small fortune as civil engineers.[16] Many West Point men found lucrative work surveying and constructing railways throughout the United States before, during, and after the Civil War.[17] Others prospered by developing the metropolis at the mouth of the Hudson. George S. Greene, a member of the class of 1823, for example, oversaw the improvement of the Croton waterworks and proposed a central underground railroad line.[18] Another case in point, Egbert Viele, returned to New York after the Mexican-American War to survey and supervise the initial construction of Central Park.[19] As will be explored in chapters 5 and 6, Viele would also go on to be a Central Park commissioner and one-term member of the U.S. House of Representatives.[20] In New York, West Point graduates could use their technical expertise as well as their political abilities to seek advancement through the expansion of the city. Slocum pursued a career in law and in Congress, representing Brooklyn at the same time the Brooklyn Bridge was being completed.[21] In nineteenth-century New York, the works of the West Point graduates and the networks formed during construction of those works became unique pieces in the urban mosaic of Gotham.

Studies of New York City's history emphasize the leadership and influence of men such as William Cullen Bryant, Fernando Wood, Frederick Law Olmsted, William "Boss" Tweed, and Andrew Haswell Green on urban development and culture in nineteenth-century New York.[22] Whether their role was technical, entrepreneurial, or political, or a combination of the three, the West Pointers who came to New York also influenced the shape and design of the nineteenth-century urban landscape. In his book *Empire City,* urban historian David Scobey argues that commercial interests and self-promotion often overshadowed civic order and virtue in Victorian New York. Scobey believes that speculation and party machine politics smashed the grand visions of Olmsted, a landscape architect and reformer, and William Martin, "a booster ideologue."[23] Instead of becoming a planned urban landscape "of civic order, capitalist dynamism, and civilized public and domestic life," New York succumbed to patterns of uneven and "unplanned" development dictated by the hard realities of the city's real estate market.[24] West Point men, with other reform-bent individuals, sought to impose order over the unwieldy forces of the city's property development, as George B. McClellan and the Dock Department did in 1872 with the proposal to build standardized stone dock works around the shoreline of Manhattan.[25] Some graduates encouraged property speculation and looked to improve their own fortunes in the process, as Viele did in the Upper West Side during the 1880s.[26] Some graduates thought they were improving order in the growing city, but inadvertently created more adversity and disorder for New Yorkers. For example, Fitz John Porter's recommendation to remove buildings in the way of Croton Aqueduct improvements in 1875 made way for the city project, but also displaced several privately owned structures. Viele displaced hundreds of African Americans and Irish immigrants who lived in Seneca Village through his clearing campaign for the building of Central Park.[27] What becomes apparent in the study of West Pointers in New York is that they were deeply drawn into the machinations of municipal and state power. Fernando Wood, "Boss" Tweed, and Andrew Green may have commanded the headlines of the day, but West Pointers' names were further down in those same news stories, which described both kinds of men in the pursuit of their goals in Gotham.

West Point historian Ted Crackel asserts that "the Academy's history is a reflection of the nation it serves, for West Point has mirrored the broader movements of American society."[28] The history of West Pointers in New York City between 1817 and 1898 provides intriguing reinforcement for Crackel's thesis, because not all the graduates who went to New York succumbed to the lure of profit, patronage, and power. The first generation of West Pointers, including George Greene and William H. Sidell, devoted their energies out of a sense of duty and moral obligation as they plied their engineering skills for the city. These early graduates of the U.S. Military Academy served in the army prior to the Mexican-American War. As civilians, they joined the movement to organize engineers into a distinct and respected profession. With their fellow civilian engineers, Greene and Sidell created the American Society of Civil Engineers, an organization that aspired to remain above political intrigue and corruption. Later graduates, like Viele, Slocum, Porter, and McClellan, would embrace politics in order to advance their plans for the city as well as their own careers, especially after the Civil War.

This book examines the experiences of Military Academy alumni in the context of New York's transformation into a Victorian metropolis. The common denominator for all the men highlighted in the following pages is the U.S. Military Academy and how the alumni's experiences, associations, and perceptions at the Academy shaped their actions and influence in the city.

Sylvanus Thayer became superintendent of the U.S. Military Academy in 1817. Often called the "Father of the Military Academy," Thayer was the first superintendent to standardize cadet training and implement the Socratic method of daily recitation in cadet education.[29] Under Thayer, the academic program of math, engineering, and very limited liberal arts became the staple of the cadet experience. Between 1831 and 1871, Dennis Hart Mahan and the academic board of the Military Academy embraced Thayer's method and made the cadet course a curriculum devoted mostly to engineering, both military and civilian. Graduates who lived and worked in nineteenth-century New York City all studied under the "Thayer method" and endured Mahan's program of engineering courses. As will be explored in chapters 2 and 3, these cadets received an education like no other in the United States to that point. In light of Thayer's and Mahan's influence, this story focuses on the classes graduating after 1821 and before 1871, when West Point gave men the ability to apply order and reason to the increasingly disordered world of nineteenth-century America.

The relationship between the American military and American society in the nineteenth century was more intertwined than in more modern times. In the early republic's embrace of technology, civilian and military innovations fueled a new economy, creating a greater emphasis on hope and optimism in the inventions of the future. Beyond the steam engine, the cotton gin, and interchangeable parts, military men possessed the expertise in math and science to build forts, bridges, and roads.[30] Indeed, they embodied the utilitarian-technological spirit that permeated the United States and was keenly observed by the Frenchman Alexis de Tocqueville.[31] The national government looked to the officers of the U.S. Army not only to defend the nation, but also to supervise the construction of canals, and later, railroads and bridges. As the United States expanded west, army topographical engineers mapped the newly occupied territories. The Army Corps of Engineers led the nation in developing national waterways, channels, and harbors across the continent. Also, army-supervised harbor and lighthouse construction projects were pivotal to the growing maritime commerce fueling the American economy. Without army engineers to guide the dredging of channels and improvement of ports, steamboat operators and shipping giants like Cornelius Vanderbilt and Samuel Cunard would have been more challenged in creating their transportation empires.[32] Engineers who had worn army blue were a vital part of commercial development in the United States from the end of the War of 1812 through the close of the century.[33]

Nowhere was this interconnectedness between the American military and civilian society more evident than in New York with West Point–educated engineers. Not only did cadets travel to New York for social engagements, but New Yorkers traveled to West Point as well. In the 1840s and 1850s, members of the New York elite spent summers escaping the stifling heat of the city by visiting West Point and observing summer encampments. George Templeton Strong's diary records many of these festive gatherings at the Cozzens Hotel on the grounds of the Military Academy.[34] One can imagine the relationships that West Point summers cultivated between these sons of duty and the elite of New York society.[35] Certainly these encounters shaped the cadets' views of New York City, but no New Yorkers could have foreseen how their world, and their city, would be shaped by the cadets who later came as engineers, professionals, war heroes, boosters, and politicians.

Why the story of these West Pointers in the city has not been explored extensively will also become evident in this book. Often they were on the losing side of the competitions that determined the course of history in Victorian New York. Viele remained in the shadow of Olmsted, Slocum under the Brooklyn machine, and McClellan will always be eclipsed by Ulysses S. Grant and Republican politics. Even so, the history of Military Academy engineers in the city creates a more complete picture of how New York was transformed in the nineteenth century. It is yet another case of New York satisfying an ambition and drive that was uniquely American.

Former military men in nineteenth-century America collectively represented another path to social status and wealth, and American society generally accepted veteran officers as having an elevated status. But this acceptance could not be fully realized until the officer left the uniformed service for civilian life. A West Point graduate who was a regular army officer had difficulty escaping the stigma of being part of an "idle" and "sinister" standing army in the nineteenth century.[36] Americans generally still harbored fears that a standing army could threaten American liberty, as the British Army had done in eighteenth-century North America. Antebellum political parties shared a distrust of standing armies, and favored citizen militia to provide the nation's defense.[37] As "civilians" in New York, West Point graduates went from the U.S. Army, an institution Americans respected but still did not yet completely trust, to a world of social and professional relationships whose meanings were generally understood by New Yorkers. So, in effect, the city transformed these West Point men as they sought to change the city. It was a dynamic relationship that defined New York's and the nation's views of West Point, especially in the decade of the Civil War. After the Civil War, the individual military experiences of the former officers had a great part in determining where and how New York society accepted them. In many cases, the more celebrated the military experience, the greater the acceptance of the officer veterans by civilians. Regardless, before, during, and after the war, city elites embraced engineering expertise garnered in and out of uniform.

The chapters that follow illuminate how this civilian acceptance of military professionals became possible through the West Pointers' influence in the city. Several chapters begin with a brief vignette highlighting a specific episode in the life of a figure central to that chapter. Where appropriate, I have included glimpses into the personal experiences of the men and women who lived in this world of professional engineers. My purpose is to make these West Pointers more than just a list of names on engineering projects, and to help the reader focus on the human dimension of these experiences. Together, Academy alumni, engineers, the city's authorities, and power brokers embarked on a journey that created the engineering profession, raised the primacy of city infrastructure and public services, and defined the scope of responsibility and skills required by urban leaders at the dawn of modern America.

1

WEST POINT INFLUENCE IN VICTORIAN GOTHAM

The student of physical science works with definite quantities, and his conclusions are based upon precise premises. In the great game of war the field is more extended, and the skillful player must combine the precision of the mathematician with the profound knowledge of the strength and weakness of human nature. Nothing can be neglected.

—HENRY ABBOTT, "Memoir of Dennis Hart Mahan," 1878

The progression from cadet to army officer to civilian was a uniquely American experience that enabled Military Academy alumni to be successful in New York. West Point's engineering curriculum, tempered by the officer experience, produced a level of expertise and high reputation that made graduate advancement in Victorian New York possible. In New York, Academy graduates influenced three important areas of metropolitan development. First, West Point graduates joined with civilian-trained engineers in the city to create the first professional organization of civil engineers in the United States, the American Society of Civil Engineers (ASCE).[1] From 1852 through the end of the century, New York City served as the home not only for ASCE but for the professional organizations created by lawyers, doctors, and architects as well.[2] Second, the infrastructure projects designed and built by West Point graduates often became the standard for urban development and public works in other nineteenth-century American cities. The innovations used in the Croton Aqueduct, for instance, were adopted in Boston, Pittsburgh, and Baltimore.[3] Third, Academy alumni defined the relationship between technically proficient professionals and the politics of Victorian Gotham. Both the leaders of Tammany Hall and reform-minded representatives in the state legislature saw West Point graduates as experts who could achieve credible results in a politically charged scene, and yet maintain a high level of respect among the general public. Tammany leaders could deny accusations of corruption by pointing to the honorable reputations of the West Point men put in charge of the municipal project or agency. In several instances, when the state legislature sought to reform municipal government, it retained or hired West Point graduates. In 1857, for example, the state-appointed Central Park Board kept Viele as engineer-in-chief of Central Park after it had relieved the city of control.[4] By the 1870s, Military Academy alumni could be counted on to serve as commissioners for the Police Department, the Streets Department, and, later, the Department of Public Works.[5] This is the story of Victorian Gotham seen through the mutual visions of city leaders, the engineering profession, and West Pointers.

PROFESSIONALISM

The remarkable effect of Military Academy officers in establishing military professionalism is well documented.[6] William Skelton sees the professionalization of the army officer corps occurring between 1815 and 1850, prior to the Civil War. According to Skelton, the American officer corps "acquired a regular system of recruitment and professional education, a well-defined area of responsibility, a considerable degree of continuity in its membership, and permanent institutions to maintain internal cohesion and military expertise." Through West Point, the westward expansion of the United States, and the government's general recognition of the army officer corps as a group with military competence, Skelton asserts, the American military was one of the first professions to emerge in the antebellum era replete with expertise, corporateness, and a sense of responsibility.[7]

Samuel Huntington's work leads the other school of thought, which argues that professionalization of the American officer corps could only occur after the Civil War.[8] Even though Huntington acknowledges that "the distinguishing characteristics of a profession as a special type of vocation are its expertise, responsibility, and corporateness," he argues that the "military profession" was unique among all vocations. Huntington holds that, in addition to these distinguishing characteristics, the "military profession" required a constructive relationship with the civilian citizenry that it defended. Central to Huntington's thesis is the concept of the "southern military tradition" whereby white southern males embraced a "cult of romantic chivalry" as they fought Native Americans, quelled the threat of slave revolts, and cultivated the prestige of the military officer as a profession. Only the antebellum South fully accepted and incorporated the military ideal into its society and institutions, making the civil-military relationship "legitimate." Citing the large number of regular officers who joined the Confederate Army at the outbreak of the Civil War, Huntington found that the North had not embraced the military profession by 1861, and thus remained "isolated from the mainstream of American development."[9] For Huntington, the military officer corps needed not only a body of expertise and competence to be professional, but it also required reciprocity of loyalty between the uniformed servicemen and the nation at large; and that did not exist prior to the Civil War.

The concept of professionalization among civilian vocations also started in the antebellum era. In his seminal monograph, Burton Bledstein finds that this identity of professionalization began in the 1840s with the emergence of a middle class. Among the first civilian vocations to organize as a profession were the civil engineers in 1852.[10] West Pointers had a profound influence on the professionalization of engineers. When considering the role that West Point–trained engineers played in forming the American Society of Civil Engineers, one can identify trends that explain why Skelton and Huntington saw the emergence of military professionalization differently.

First, Huntington fails to grasp the nexus between the army and civil engineering. According to Skelton, antebellum engineer officers did not resign as soon as they had fulfilled their service obligations. To the contrary, his analysis of the army registers between 1830 and 1860 showed that more than half stayed in uniform for twenty years and another 40 percent remained for thirty. Of course, Skelton's study was not limited to West Point–trained veterans, as it took into account all army officers serving in the antebellum period. Even with the majority of military engineers making their careers in the army, Skelton acknowledges that the Military Academy engineers did play "a central role in shaping the American civil engineering profession."[11]

Second, the difference between these two views of antebellum military professionalism depends on where one places the centrality of loyalty in defining the military profession, and professions in general. Civilian professionals tended to emphasize loyalty to the profession over loyalty to a political group or party.[12] Graduates of the antebellum Academy had developed a sense of loyalty to the nation, but their education focused on following military science and engineering principles taught by the school. Regardless of the actual causes of the defection of one-third of West Point graduates to the Confederacy, the American public perceived the cause to be a shortcoming of the national Military Academy's system for producing military officers. Civil War veterans on both sides spent the rest of their lives working to reconcile those differences that had fueled their animosity during the war. It took the Military Academy three decades to reconcile with southern graduates who had defected to the Confederacy.[13] Eventually, loyalty to the nation became paramount—in the 1890s, West Point adopted "Duty Honor Country" as its official motto.[14] Thus, the military profession diverged from the prevailing concept of professionalization as understood by doctors, lawyers, engineers, and architects. Unlike civilian callings, the military officer at the end of the nineteenth century was expected to be a loyal patriot and a master of military expertise, and to render responsible service to the nation's defense.

In broad terms, Bledstein's study of professionalism in the civilian realm and Skelton's description of antebellum military professionalization are similar responses to the social changes occurring in nineteenth-century America. This book examines the rise of professionalization through West Pointers in New York, and concurs with Skelton's views of military professionalism. Consideration of the graduates' actions in both military and engineering professionalism provides a more useful understanding of professionalism overall as it emerged in nineteenth-century America, specifically in New York City.

INTERNAL IMPROVEMENTS

While West Point men played a well-known role in the internal improvements of the United States, their roles in the internal improvements of New York often proved to be more lucrative and professionally rewarding.[15] The phrase "internal improvements" when applied to nineteenth-century America normally pertains to canals, roads, bridges, railroads, and public works—the infrastructure—that were funded by the federal government and intended to benefit two or more states. For the purposes of this book, public works, public buildings, parks, sewers, and street construction will be treated as "internal improvements" for New York, especially in those instances where state funds financed construction projects in the city. The concepts of "internal improvements" and "public works" are similar in that they created a relationship between the Academy graduate and the public funding of the project employing their expertise. Like those West Point alumni who profited from working on federally funded national improvements, West Point alumni in New York profited from the publicly funded development of nineteenth-century Gotham.

The making of modern New York City was an important model for other nineteenth-century American cities. John Randel's grid street system as well as local municipal railroads first started in New York, and could later be found in newer American cities such as Oklahoma City and Salt Lake City.[16] Manhattan's patterned street plan expressed the desire of the early American republic to master the unpredictability of nature, both figuratively and literally. City leaders wrestled with competing tensions between manmade progress and the preservation of nature and all that it inspired.[17] Engineering projects, built for functional outcomes, also symbolized the progress of the brave new republic. When completed in 1842, the Croton Aqueduct with its High Bridge over the Harlem River was an engineering feat celebrated across the country and in Europe.[18] Central Park was the standard for publicly created green space in American cities.[19] Engineers in William "Boss" Tweed's Department of Public Works demonstrated the benefits of centralized, comprehensive sewer planning and construction in the 1870s.[20] Washington Roebling's techniques used to build the towers and suspension cables of the Brooklyn Bridge were used on other bridge projects through the first decades of the twentieth century.[21] New York stimulated the masses, almost to a fault. Whether they celebrated the city or abhorred its excesses, nineteenth-century Americans had to recognize New York as the nation's leading and most influential metropolis.[22] James Fenimore Cooper astutely observed in 1851 that "New York is essentially national in interests, position, and pursuits. No one thinks of the place as belonging to a particular State, but to the United States."[23] New York was the "national city" to which Americans compared other cities, domestically and abroad, engineering feats included.

Central to understanding Gotham's development in the Victorian era is the research of David Scobey. He places the "transformation" of New York in the "third quarter of the nineteenth century, some thirty years earlier than the usual periodization of morphological change." Instead of focusing on technological progress in the late nineteenth century as the agent of change, Scobey treats the development of the "metropolitan real-estate economy as the key

engine of change." Scobey argues that "the mix of dynamism and market frag-
mentation that characterized city building in the midcentury boom" forced
"propertied New Yorkers . . . to come together in new ways to manage the jug-
gernaut of change." Looking at "[trade] journals like the *Real Estate Record and
Builders' Guide,* developers' coalitions like the West Side Association, [and] reform groups like the Citizens' Association," Scobey describes how the development of the city mixed the classes and networks in New York. As a result of the
real estate commodity forces, "Victorian Manhattan" was "part of an emergent
pattern of uneven development" resulting from collaboration and interaction
of speculators, politicians, professional engineers, and working-class groups.[24]
Between 1840 and 1870, "intellectuals, engineers, sanitarians, and design practitioners" pursued the ideal of William Cullen Bryant's vision of an ordered
"shaping of urban space and urban growth."[25] In New York, this drive for or-
derly urbanization was just as much an application of technical expertise as it
was a deliberate effort to create "social and moral improvement."[26] So, in the
physical changing of the urban landscape, the forces of social change, technical
innovation, and political power all acted together to create Victorian New York.

Scobey examines the period of New York's development when West Point
men arrived on the scene. Beginning in 1856, Egbert Viele entered this urban
development as an engineer, politician, booster, and socialite. His professional
life is illustrative of the city Scobey describes. Initially, Viele was an engineer
spearheading Central Park's construction, and over the next five decades his
influence and activities blurred the lines between "professional" and "politician." Sometimes, he was a virtuous reformer seeking to end the public hazards
of "miasmatic odors" through a public works sanitation campaign.[27] At other
times, he was a booster and West Side development lobbyist campaigning for
the state to fund the improvement of Riverside Drive.[28] For this story (as well
as the subject of chapter 5), Viele is a bridging figure who connects the identity
of the West Pointer as a professional with the identity of the West Pointer as
politician. Viele was not as successful as he aspired to be in either career, but he
was in New York for nearly sixty years advocating his vision of the metropolis
and its culture, not to mention trying to aggrandize his legacy in the process.[29]

George McClellan's Civil War record makes him a more recognizable example of a West Point graduate in the city. McClellan spent his time in New
York mainly in the political arena, beginning with the Democratic Party nomination as its 1864 presidential candidate. From New York, he also pursued

business interests elsewhere. After losing the election to Lincoln, McClellan started an engineering consultant business and lobbied to become president of New Jersey's Morris and Essex Railroad. According to one of McClellan's patrons, industrialist and New York mayor Abram Hewitt, the board of directors was not comfortable with the defeated candidate as its company president for fear that it would jeopardize company relations with the government. McClellan set off to tour Europe in January 1865 in a self-imposed exile that lasted three years, only to return to New York for another possible presidential campaign in late 1868. As will be seen in chapter 6, McClellan walked a delicate line between the political realm and that of "professional engineer." Careful to remain free of any stigma that could come from associating with Tweed and the Tammany Democrats, McClellan accepted a position as head of New York's Department of Docks, even though more important and influential posts had been offered to him. The key for McClellan and other West Point graduates in New York was not to compromise or diminish their reputations built upon the Military Academy, their war records, and their corporate expertise in the engineering profession. By declining the job of city comptroller in 1871, McClellan was able to preserve and, perhaps, rehabilitate his reputation enough to become governor of New Jersey in 1878. Even though McClellan ultimately attained his highest elected office in New Jersey, he had earned the Democratic Party's trust and restored his reputation while he served as a public docks' official in New York.[30]

As both Viele's and McClellan's parts in New York's urban development suggest, merely being a proficient or well-connected engineer was not enough to be a successful professional engineer in Victorian New York. Internal improvements resulted from a confluence of technical innovation, municipal power derived from socioeconomic networks, and political power. Sven Beckert aptly describes the era in *The Monied Metropolis: New York and the Consolidation of the American Bourgeoisie, 1850–1896*. His book explores how "capital-owning New Yorkers overcame their distinct antebellum identities, rooted in the ownership of different kinds of capital, to forge in the wake of the Civil War dense social networks, to create powerful social institutions, and to articulate an increasingly coherent view of the world and their place within it."[31] When West Point graduates came to New York to apply their engineering expertise in the improvement of the city, the more notable ones became part of the "social networks" and "powerful social institutions" that Beckert examines.

In seeking to build New York, West Point alumni also used their expertise and reputations to enter the city's powerful elite class of capitalist businessmen and professionals.

MUNICIPAL POLITICS AND ADMINISTRATION

While men like McClellan and Viele used their engineering expertise as a means to access those with political influence in the city, others like Henry Warner Slocum relied on their war record to focus their postwar career on the municipal political scene. As a lawyer and congressman in Brooklyn, Slocum possessed war hero status and a reputation for integrity.[32] Slocum and Viele would be part of the debates over the development of Brooklyn, specifically the Brooklyn Bridge and Prospect Park. Greed and patronage often masked the desired outcomes of reform-minded professionals such as Frederick Law Olmsted and William Martin, head of the West Side Association. Viele, a Democrat, spent a public career opposing Olmsted, a Republican, in the creation and management of Central Park and Brooklyn's Prospect Park, but he would serve with Martin as an active member of the West Side Association.[33] Slocum, also a Democrat, sought to remain above the Brooklyn political machine of Hugh McLaughlin. In some cases, Military Academy graduates were against reform initiatives or campaigns, but more often than not, these former military men were for reforming the ills created by forces of change in nineteenth-century New York. Conflicts among reformers usually emanated from the different visions of how to achieve reform in Gotham's social and political arenas.[34] Later chapters will explore these relationships and consider how West Point men aligned their political, professional, and social loyalties.

New York would look to West Point men to serve as commissioners to manage city works and public services. As municipal commissioners, they could attempt to distance themselves from the contact sport that New York politics became after the Civil War. As streets commissioners, police commissioners, and even docks commissioner, West Point–educated men could leverage their reputation and professionalism in the city's transformation and advance their own position in society. Fitz John Porter desperately desired to serve in national government after the Civil War, but his court-martial conviction for failing to follow orders at Second Manassas confined him to accepting municipal appointments. William Farrar Smith served as a police commissioner in an age when the New York police were viewed as being as corrupt as City Hall permitted. After forty-eight years of service in the army's Corps of Engineers, John Newton became the commissioner of public works in 1886.[35] These individuals are just a few of the West Point men whom the City of New York appointed to run the metropolis after the Civil War. The selection of West Point alumni to serve in various commissioner appointments was a natural development in the relationship between the city's elite and the military men. For nearly eight decades, the Thayer method and Mahan's engineering program had produced a group of engineers and former military men who were technically proficient, politically astute, and professional. What is more, after the Civil War, these individuals were patriots whose heroism in battle, real or perceived, could be exploited by the associations, organizations, and enterprises to which they belonged. For good or for ill, Academy alumni were an integral part of the reform narrative that ran through New York City's history in the last decades of the nineteenth century.

Although many West Pointers who came to New York had fought in the Civil War for the Union and the end of slavery in the United States, they often did not continue to support the cause of black Americans and freedmen in postwar New York. As Democrats, they resisted federal efforts to advance the freedoms and rights of the black population in the era of Reconstruction.[36] There were limits to what could be achieved by the West Pointers' professionalism and approach to improving the city. New York would be a mixture of old and new immigrants at the turn of the century, but the racial order would remain much as it had been at the turn of the nineteenth century.

By the end of the century, New York had evolved into Greater New York, consolidating the five boroughs of the Bronx, Manhattan, Queens, Staten Island, and Brooklyn on New Year's Day 1898. This process was not completed by any one means but, rather, Gotham merged the five boroughs through a variety of political, social, financial, and physical developments. The consolidation of Greater New York resulted from the confluence of progressive visions of reform, the pursuit of municipal efficiencies, and a campaign to remain the leading "imperial city" in a time marked by the emerging national impulse to be imperial.[37]

Connections between the West Pointers and Gotham's leading figures in the decades preceding this municipal transformation created a network of power and expertise that shaped, influenced, and ultimately transformed New York into the center of the nation. This transformation has its roots in ante-bellum West Point and the graduates who embraced the Military Academy's fundamental approach to science, technology, and engineering as leaders in nineteenth-century America.

2

AN AMERICAN ECOLE POLYTECHNIQUE

The Science of an Engineer is applicable to almost every profession in Life; it is highly essential in some and injurious to none.
—MAJ. JONATHAN WILLIAMS to Maj. Decius Wadsworth, 13 August 1802

The June morning had begun at sunrise with a cannon shot, a military parade, and breakfast, all with the pomp and precision expected at the United States Military Academy in 1824.[1] At 8:30, twenty-two first-class cadets stood ready for several hours of questioning before the superintendent, the Academic Board, and the Board of Visitors, who were assembled in the large examination room. At one of the two enormous blackboards up front was Cadet Dennis Hart Mahan, who was all too familiar to the men of the Academic Board, and to the school's superintendent, Colonel Sylvanus Thayer. The young Mahan, while still a cadet, had served as an acting assistant professor of mathematics for the previous three years. For Mahan, this was just another routine examination in mathematics and civil engineering. Lieutenant Courtenay, himself a West Point graduate of just three years prior, took each cadet through the paces of the examination.[2] No doubt the heat and humidity of the Hudson Valley in the early summer contributed to the stress and the sweat of the cadets dressed in their woolen gray uniforms. After four and a half hours, Mahan emerged from the examination room as the number one cadet in mathematics and engineering. For the remainder of June, Mahan and the rest of the Corps of Cadets would repeat this process for all their courses in the Academy curriculum, a curriculum greatly enhanced and standardized after Thayer's arrival in 1817. By graduation day on 1 July 1824, only Mahan, the number one cadet of the class overall, scored well enough to earn a commission as second lieutenant in the prestigious Corps of Engineers, and to be asked to serve as a member of the faculty.[3] While Mahan proved to be an exceptional graduate of the Academy in 1824, in the antebellum era all West Point graduates completed Thayer's program of study with this rite of passage to become officers in the United States Army.

Between 1817 and 1871, some 2,249 cadets graduated from the U.S. Military Academy at West Point. These graduates all shared in the "West Point experience" highlighted by their reception as new cadets, mastery of a program of study centered on science and engineering, drilling on the plain, and the "Thayer method" of instruction. Out of this common experience, West Point men not only served as officers in the army but also ventured into a variety of enterprises and disciplines. During the American Civil War, most West Point–educated men defended the Union, while a substantial number fought to undo it. Before and after the Civil War, Academy graduates directed numerous public and private works projects. Throughout this period, many shaped the professionalization of the engineering field in the United States. Regardless of where or how the graduate made his way after his time at the Academy, all Academy graduates became part of the West Point contribution to the history of the country. And common to every graduate's experience at West Point was the curriculum and program of cadet development at the U.S. Military Academy.

When Sylvanus Thayer arrived as the superintendent in 1817, he set about standardizing the curriculum and shaping the Academy's program to match that of the Ecole Polytechnique, the military and engineering school created under Napoleon in France. From 1817 to 1833, Thayer transformed West Point into a professional engineering school that produced military officers ready for service in the U.S. Army.

Not as celebrated, but probably just as significant to the antebellum Academy, was the tenure of Dennis Hart Mahan. After initial duties as an instructor at West Point, Mahan, like Thayer, studied math and engineering in France. In the summer of 1830, Mahan returned from Europe as a newly appointed assistant professor of engineering. He spent four decades teaching cadets military and civil engineering as well as military science. From 1830 to 1871, Mahan was the first and major influence on West Point graduates in military science, fortifications, and civil engineering. Every cadet who graduated from West Point during Mahan's tenure had to take at least one of his courses.[4] First as a cadet, then as a lieutenant, and finally as a civilian professor, Mahan embraced

Thayer's program of study and instruction. Thus, Thayer's methods and curriculum were greatly improved by Professor Mahan's program of engineering and military science. He sought to ensure that the army's leaders from West Point could fortify the nation as well as lead American soldiers in battle.[5]

Since the founding of West Point in 1802, there has been debate as to what sort of education a military institution should provide for the future officers of a nation's army, especially the army of a democratic republic. How should an institution create a group of leaders both virtuous and expert? How can men be made who are capable of moving the nation forward? Should the military school's curriculum include only military training, or should it prepare leaders for building fortifications, firing artillery, and applying other sciences? In the first decade of the twenty-first century, in places like Kabul, Afghanistan, where West Point graduates and faculty have helped to establish the new National Military Academy of Afghanistan, the scope and purpose of the Military Academy's current curriculum have prompted the same discussion that surrounded West Point some two centuries prior.[6] American leaders in recent decades have tried to avoid "nation building" with the American military.[7] But to advocate such a policy is to ignore the actual experience of the American army, particularly the experience of the American army led by West Point graduates in the nineteenth century. From its beginning, the United States has looked to West Point men trained in engineering to build the roads, canals, and railways that have led the way in making the new republic. In the Age of Jackson, when critics of the Military Academy portrayed it as a federally funded institution of antidemocratic elitism, West Point's proponents defended the benefits of the science-and-math-based curriculum for the nation.[8] By the mid-nineteenth century, even America's university leaders publicly acknowledged the contributions of Academy graduates in the construction of American railroads.[9] Whether in 1802 or at present, the nature and scope of a military education have had implications well beyond the realm of fighting a nation's wars. In the history of the United States and its military academy, antebellum demands set the precedent for an engineering and scientifically oriented educational military institution in America.

West Point has always been an institution dedicated first and foremost to producing military leaders to serve the needs of the nation. The military accomplishments of West Point graduates permeate almost every aspect of nineteenth-century American military history, from the War of 1812 through the Spanish-American War. As critical as the West Pointers' military accomplishments were to American success in war, their contributions to the physical, political, and social changes of the nation were just as important to the growth of the United States. The graduates of West Point capitalized on their unique education in the making of America, especially in the years following the formalization of the scientific curriculum under Thayer.[10] James Morrison argues that "the institution . . . left its mark on the American Experience" and conversely, that "American History has molded West Point."[11] Ted Crackel echoes this assessment: "The Academy's history is a reflection of the nation it serves, for West Point has mirrored the broader movements of American Society."[12] The men educated at West Point who led American soldiers in battle also designed, supported, and built the infrastructure for nineteenth-century America.

THE U.S. MILITARY ACADEMY CURRICULUM
IN ANTEBELLUM AMERICA

Plainly, the early curriculum of West Point was distinct from that at other American colleges and universities in the early nineteenth century. For nearly four decades prior to the Civil War, the U.S. Military Academy was the leading institution of engineering in the United States. From its first year as the national military academy, West Point was to be a unique institution of higher learning in the United States. Maj. Jonathan Williams, the first superintendent, wrote, "We must always have it in view that our Officers are to be men of Science, and as such will by their acquirements be entitled to the notice of learned societies."[13] The study of science became the method for creating a professional class of men who could lead a standing American army and yet still remain above party, factional, or sectional loyalties. As Thayer set forth to make the Military Academy a scientific school, he was going against the prevailing wisdom of antebellum college curricula. The majority of American colleges held fast to the traditional curriculum of the age, concentrating on the classics, mathematics, and moral philosophy. For example, Yale College's faculty issued a report in 1828 defending their resistance to the radicalism of a scientific curriculum. In Thayer's view, the Board of Visitors, which served as West Point's "board of trustees," demanded that he establish and perfect a practical system of study for the cadets.[14] Moreover, the national mission of the Academy obligated West Point to prepare and educate the cadets for duty as leaders in the American army.

The emphasis on science, mathematics, and engineering determined the nature of American security strategy between 1802 and 1865. Historian Brian Linn provides an explanation of how early nineteenth-century strategic and military thinking correlated with an ever-increasing emphasis on an engineering curriculum. Specifically, in the era after the War of 1812, civilian and military leaders surmised that the greatest threat to the young republic would come from an attack on the coast by a political or economic rival from Europe.[15] As a result, the national leadership devised the requirement for educated engineers to design and oversee the construction of coastal fortifications. Thus, West Point had the duty to educate the Corps of Cadets to be those engineers. In this sense, the perceived need for engineers to build "Fortress America" led to the demand for West Point–educated engineer officers, and conversely, West Point–trained engineers propagated the "Fortress America" strategy, even when threats to national security indicated otherwise. When the United States went to war against Mexico in 1847, the army's leadership, led by engineers, realized that officers needed to know more about leading a campaign march than, say, building a fort on the coast. The Mexican War and subsequent frontier duties drove many West Point–educated officers to resign their commissions and find other means to achieve success, and for many, to use their scientifically based education. In the years between the Mexican War and the Civil War, the Board of Visitors and the Academic Board did pursue minor adjustments to West Point's curriculum. One of the better-known adjustments was the attempt to include more military science by adding a fifth year to the course of study between 1854 and 1861. Military science was really the study of military history. At West Point, Mahan's textbook, *An Elementary Treatise on Advanced Guard, Out Post and Detachment Service of Troops,* was the main text on military science for the cadets.[16] However, West Point's curriculum remained focused on science and engineering and was largely unchanged through the end of the Civil War. To be sure, though, the Mexican-American War legitimized the need for a "professional officer corps" and ended any discussion about "abolishing the military academy at West Point" up to that point.[17]

The nation and the Academy formed a relationship of mutual expectations in which the school would supply military officers trained in engineering, and the nation would put them to work using those skills. Additionally, the United States expected West Point to produce patriotic men loyal to the national government. As the Civil War proved by the instances of graduates who defected to the Confederacy, it was easier for the Military Academy to produce engineers than to forge a loyal homogenous officer corps. Critics of the Academy pointed to the school's overemphasis on the engineering curriculum as one of the reasons graduates left the Union for the South. But others argued later that there was no way in which the Military Academy could overcome the sectional differences present in antebellum America. The issue of loyalty as a requisite to become a professional military engineer could not be settled until the entire country decided the question of national loyalty during the Civil War.[18]

In the study of the Military Academy's education prior to 1861, many military historians have looked to connect the cadet experience with the conduct of the war by graduates on both sides of the Civil War.[19] The indictment of the U.S. Military Academy's role in contributing to the national schism of 1861 generally follows two lines of logic. The first, and perhaps more circumstantial, argument asserts that the Academy failed to overcome sectional disparities in the Corps of Cadets, and therefore undermined the spirit of a "national academy" with cadets nominated from every congressional district in the Union. A second line of indictment argues that Mahan's military science course and the emphasis of the French military theorist Antoine-Henri Jomini in the cadet military curriculum led commanders on both sides of the war to pursue bloody, large-scale battles. In effect, Mahan's tutelage was responsible for the national calamity of casualties between 1861 and 1865.[20] This second argument is more debatable, especially since the most influential commanders on both sides had at least five years of military service after graduation to inform their military judgment and decisions in battle. Regardless of the debate over national loyalty and military professionalism, the leaders of New York had no qualms about accepting West Point engineers for their civil engineering expertise before and after the Civil War.

MILITARY PROFESSIONALISM

The U.S. Military Academy and the U.S. Army both worked to define what a professional military officer should be in the nineteenth century. Recall that the history of the military professionalism debate falls into two basic camps as described in chapter 1. To recap, the first, best espoused by William Skelton in *An American Profession of Arms: The Army Officer Corps, 1784–1861*, sees the professionalization of the army's officer corps occurring between the War of 1812 and the Civil War. The second, led by Samuel Huntington in *The Soldier and the*

State, argues that the American officer corps did not become truly professional until after the Civil War, when the loyalty of the military to the nation was no longer an issue.[21]

Marcus Cunliffe's work helps frame the emergence of military professionalism in the United States. Cunliffe supports Huntington's characterization of the martial spirit in the prewar South, but takes issue with the idea of a schism between military officers and antebellum northerners. Moreover, Cunliffe's research reveals more myth than reality to the "militaristic South," as he found a higher number of incidents of mob violence and unruliness in the antebellum North than in the South. Granted that the southern military academies, specifically the Virginia Military Institute and the Citadel, achieved greater popularity and prestige in southern society than military schools achieved in the North, but the idea of the military school began in the North in 1819 with Norwich Military Academy and led to military academies in the Northeast. Thus, the military profession had at least a presence in, if not a relationship with, all sections of the antebellum United States. Additionally, Cunliffe argues that prior to 1860, "[prospects] for the officers were so limited that many resigned their commissions at the earliest opportunity. Whether they stayed in the army or left it, American regular officers retained civilian characteristics to a considerable degree." The differences between military officers and civilian gentlemen were not as pronounced in the nineteenth century as they became in the twentieth. Between 1825 and 1860, West Point's prestige increased substantially as its graduates found greater social acceptance and academic respect. In support of Skelton's argument, Cunliffe finds that "the army was . . . professionalized at the same time that it appeared to become increasingly civilianized."[22] The antebellum officer corps was professional and was viewed as such.

More recently, Samuel Watson has argued that the antebellum officer corps, especially Regular Army officers, maintained a "responsibility to the nation as a whole, a responsibility conveyed through subordination and accountability to the nation." Although much of the army's attention was given to the frontier, specifically preventing Indian attacks on white settlers, during the 1850s the officers turned their professional development eastward toward the lessons of the Napoleonic wars. They viewed frontier duty as an aberration and believed that the true professional continued to study the military examples of the great European powers. Watson submits that the political affiliations of Regular Army officers were rooted more in conservatism than in sectionalism. Even during Jefferson Davis's tenure as secretary of war (1853–1857), when he arguably favored admitting more southern cadets into West Point, "[the] vast majority of officers continued to come from West Point, with its socialization in statism (an affinity for powerful government), nationalism, and professionalism."

In sum, Watson finds that military professionalism emerged in three waves. Prior to the Civil War, the first wave of military professionalism was founded on "commitment, cohesion, and responsibility." The second and third waves came after the war, with one based on "conventional expertise" and the other on "mission," in which the officer corps focused its expertise on tasks assigned by the national command authority.[23] Essentially, the development of military professionalism evolved over the course of the nineteenth century and was first identifiable in the decade before the Civil War.

The Military Academy strived to produce leaders who were loyal to the nation as well as proficient in their military expertise. As designed in 1802, West Point was to produce national unity based upon the "patriotic values" of its graduates. The Military Academy failed to achieve a cohesive sense of "patriotic values" by not erasing sectional allegiances in its graduates prior to 1861.[24] However, the school did create an emerging identity of the professional military officer, complete with a sense of corporateness, body of expertise, and sense of responsibility. During that same period, "an aspiring middle class in America was beginning to build a professional foundation for an institutional order, a foundation in universal, scientific, and predictable principles."[25] So, both the military and civilian ideas of professionalism emerged in the 1840s, with neither directly addressing patriotism nor loyalty to the nation. After the Civil War, both West Point and the U.S. Army incorporated loyalty to the nation as a key component in becoming a military officer—loyalty nourished by the cadet experience.

A more comprehensive approach to evaluating the correlation between cadet education at West Point and the battlefield leadership of graduates considers all variables of experience starting from the Academy and up to the war. In that sense, then, not only do military teachings and principles have an effect in shaping the graduates' careers and actions, but also the whole curriculum centered on science and engineering has an effect. Given West Point's curriculum, the logic and adherence to "rules" during the conduct of leading men in battle makes some sense. If war is a problem, then science was seen as a way to overcome that problem. Mahan wrote, "[The] Military Art, in all its branches, is founded upon a comprehensive and thorough knowledge of the exact and physical sciences; and in no one branch is the importance of this knowledge more felt than in Engineering."[26] Engineering provided not just a practical skill for building fortifications and bridges; it also provided an apparatus or approach to solving problems.[27] In New York, graduates would use engineering to solve the urban problems of the Victorian metropolis.

ENGINEERING EDUCATION
IN NINETEENTH-CENTURY AMERICA

The U.S. Military Academy was the first school of higher education in America to award degrees based upon science and engineering, specifically after 1817 and at Thayer's direction. More utilitarian than abstract, a West Point education distinctly followed the tradition of the Ecole Militaire and the Ecole Polytechnique in early nineteenth-century France. Conversely, civilian colleges in early nineteenth-century America modeled the elite schools of England, centering on a classical education of "arts and letters" as well as the humanities. Civilian college students studied Latin and Greek, with almost no math or science. Another difference was that West Point did not require any proficiency in a foreign language for admission.[28] Upon admission into the Corps of Cadets, cadets were taught French in order to facilitate the reading and comprehension of the many French engineering and military texts obtained by early West Point instructors, chiefly Thayer and Mahan.

Through West Point, the French tradition of civil engineering migrated to the United States and combined with the ideals of French Freemasonry to place the theories and practice of science and math at the center of nineteenth-century academia. More than just experts in the principles of math and science, engineers trained in the tradition of the Ecole Polytechnique promised a "technological transcendence" for civilization. According to French Mason Auguste Comte, a graduate of the Ecole Polytechnique, engineers could apply theory in practical endeavors—they made theoretical concepts real. The result, Comte advocated, would be "a new social order" and the foundation for his positivist beliefs. Technology consisted of the instruments and machinery that engineers would use to create this new social order.[29] In early nineteenth-century America, technology pointed the young democratic nation to a unique place, distinct from France and the rest of Europe, in the progress of science and society.[30] Engineers had to be the foundation of this new movement, and that was the purpose of the Military Academy. The Ecole Polytechnique drew students from across the population based upon merit and aptitude rather than aristocratic privilege or social class. Thus, Thayer admired this example of the French school and desired to emulate it at West Point.[31]

The education of the cadets had one object, and Thayer built the program of study to advance it. In 1826, Thayer wrote, "At many other Colleges of the Country . . . the students learn little more than the Technical Terms of the Sciences . . . [at West Point] the students are allowed sufficient time to make themselves thoroughly acquainted with the principles and practical application of each [science]. This is the real secret of that proficiency which has elicited the admiration and applause of all who have witnessed our Examinations."[32] The Academy examinations extolled by Thayer and his supporters directly represented the practical nature of the engineering curriculum designed to turn cadets into professional soldiers. Prior to the 1840s, civilian schools did include capstone courses taught to seniors by the head of the college, but they were not nearly as functional and practical as the Academy's courses.[33]

Like their counterparts in Europe, American universities still regarded engineering as the work of artisans and craftsmen, and they did not include the discipline in their curricula until the 1830s and 1840s. American college elites of the early nineteenth century looked down upon the practical sciences such as the engineering curriculum advocated by Thayer and the Military Academy. Engineering was seen as just another technical skill to be learned by artisans and craftsmen. Similar to other artisan and mechanical trades, antebellum engineering matured from a system of master and apprentice into a more formalized establishment of trade and scientific schools. Whether started by businessmen or philanthropists, these "institutes" appeared initially in the Northeast and taught engineering techniques and practices. They published books as well as newspaper and magazine articles, and sponsored meetings and lectures to share ideas and lessons learned through the 1840s. By the 1850s, though, these institutes and publications became more interested in celebrating the latest inventions and less focused on a disciplined study of engineering and academic inquiry.[34]

Still, there were other efforts to bring engineering and the study of scientific disciplines into American higher education. Some liberal arts colleges did try to incorporate engineering courses into their curriculum. However, the traditional bachelor of arts curriculum remained the main focus of colleges as they continued to award only BA degrees.[35] Early polytechnic schools went further, created full engineering programs, and awarded engineering degrees. Rensselaer Polytechnic Institute at Troy, New York, was one of the most notable new polytechnic schools. Established in 1824, Rensselaer's program was only a one-year term, and engineering was not its main focus. In fact, Rensselaer did not award its first civil engineering degree until 1835.[36] Norwich University, founded in 1819 by Thayer's rival Alden Partridge, started teaching engineering in 1821 and awarded its first engineering degree in 1824. Thomas Jefferson's University of Virginia, created in 1825, introduced its first course in engineering in 1833 and then a School of Civil Engineering in 1835. Not until the mid-1840s and 1850s did other colleges and universities establish engineering departments. Union College in 1845, Harvard in 1847, Dartmouth in 1851, Yale in 1852, and the University of Michigan in 1855 are all examples. The Massachusetts Institute of Technology did not have a department of engineering until 1865.[37]

After the Morrill Act passed in 1862, colleges and universities embraced full engineering programs over "partial" or "select course" engineering curricula. Officially called the Morrill Land Grant College Act, the bill transferred 30,000 acres of federal land to the states based upon each state's congressional delegation numbers. The states sold the land and used the funds to build new schools of higher learning with the main requirement being that they include study of "agriculture and the mechanic arts." Many of the new institutions created engineering schools. By 1872, the number of engineering schools in the United States had increased from six to twenty. Unlike the Academy, these new institutions also created graduate programs in engineering and quickly passed West Point as the epitome of engineering education in America.[38]

Central to the expansion of engineering schools during the pre–Civil War period was the number of West Point graduates who led the new programs. Harvard, Yale, the University of Michigan, and Columbia University all established engineering programs headed by men educated at West Point.[39] In some respects, West Point's success in the antebellum era led to its decline as the engineering leader in the United States after the Civil War. While the new civilian schools initiated engineering programs employing a mix of West Point alumni and other scholars, the faculty and program at West Point remained static. Lifelong appointments of Academy graduates to West Point's academic posts "resulted in the gradual formation of reactionary policies with respect to the curriculum and teaching methods of West Point." Other engineering schools in America routinely hired new faculty with new ideas, whereas "the Academy stagnated at the level of 1833."[40] As stated earlier, prior to the Civil

War, West Point was the only institution with a fully developed undergraduate engineering program grounded in science and math, though things would rapidly change.[41]

Originating as an idea during the American Revolutionary War, the military academy established by Thomas Jefferson produced not only military leaders and engineers, but a cadre of civilian engineers trained in science and math.[42] While some might argue that Jefferson saw West Point as a kind of national academy of "scientific learning," the reality is that the Academy grew into that type of institution over the course of several decades, especially after the Thayer years, and more specifically during Mahan's tenure as a professor of engineering and military science.[43] Regardless of the political and social forces seeking to steer the direction of the nation's Military Academy, nineteenth-century West Point consistently remained dedicated to the disciplines of science, math, and engineering—all necessary for building the nascent nation and realizing the potential of its rich interior.

3

THE ACADEMY OF
THAYER AND MAHAN

Our Alma Mater had done a good work, and the nation is proud of her, or ought to be, but this should not blind us to her shortcomings if any there be, or dampen our zeal to make her still more useful, and beautiful, till she shall become the beau ideal I have dreamed of for half a century.

—SYLVANUS THAYER to Robert Anderson, 12 February 1869

Together, Sylvanus Thayer and Dennis Hart Mahan had the most enduring influence on the West Point experience of the antebellum officer corps. The importance of Thayer and Mahan in the education of the cadets became evident as graduates pursued endeavors out of the army uniform. In his book *"The Best School": West Point, 1833–1866,* James Morrison Jr. notes that between 1833 and 1866, 24 percent of West Point graduates left the army to enter practice as an engineer, and 12 percent became college professors, mainly teaching mathematics, engineering, experimental and natural philosophy, and military science.[1] If Thayer was the "Father of the Military Academy," then Mahan was undoubtedly his responsible heir, faithfully shepherding cadets through the engineering curriculum.

Evidence of his perseverance and character, Sylvanus Thayer served as superintendent of the U.S. Military Academy through the administrations of Presidents James Monroe, John Quincy Adams, and Andrew Jackson. Under Monroe and Adams, Thayer used the presidents' support to focus the curriculum on math and engineering and, ultimately, to fulfill his vision of what the Military Academy should be teaching its cadets.

Prior to his appointment as superintendent, Thayer traveled to France to procure textbooks on military fortifications and engineering. He arrived in the summer of 1815 after the final defeat of Napoleon's France by European allied forces. Thayer and his traveling companion, Brevet Lieutenant Colonel William McRee, witnessed the occupation and sacking of Paris by the allies. In this period of turmoil and unrest in post-Napoleonic France, Thayer and McRee obtained for the Academy what became its core collection of French texts in math and engineering.[2] As noted in chapter 2, American engineers, especially the men at West Point, regarded French engineering as the best in the world. If West Point was to be effective in educating military officers in engineering, imitating the French system was the most logical choice in the first two decades of the nineteenth century. Education at the Ecole Polytechnique and associated schools in Paris had an immediate and very real outcome. By emphasizing that valid knowledge comes from experience and scientific observation, the method advocated by Auguste Comte and the positivists, the French schools produced the corps of Napoleon's officers who achieved legendary success on European battlefields.[3] The Ecole Polytechnique's impact began during Napoleon's rise and continued well into nineteenth-century France. Between 1795 and 1835, only 5,502 of 14,000 candidates were accepted, and the graduates occupied some of the highest government positions and offices in postrevolutionary France.[4] The Ecole Polytechnique was the shining example of what a national military educational institution should be. Thayer was not only intrigued but took to heart all he saw and experienced in Paris. Upon his return to West Point, he committed himself to adapting the French system of engineering and military science education to the curriculum of the U.S. Military Academy.

West Point's shift toward modeling the French program of education had begun while Thayer was abroad. In 1816, when the Military Academy was drifting under Alden Partridge's controversial direction, Colonel Joseph Swift, chief of the Corps of Engineers, appointed Frenchman Claude Crozet to teach engineering at West Point.[5] Partridge had maneuvered himself to be appointed the first independent superintendent of the Academy, while Swift focused his attention on Corps of Engineers matters.[6] Crozet, an 1807 graduate of the Ecole Polytechnique, had served in Napoleon's army, where he was a bridge engineer during the Battle of Wagram and the occupation of Holland and Germany. After two years as a prisoner of war in Russia, and Napoleon's Hundred Days in 1815, Crozet was tired of war, and more than likely unhappy with the state of affairs in France. In 1816, he and his wife sailed for New York, where they met Swift, and he accepted Swift's offer to teach at West Point.[7] Charged with teaching chemistry, natural philosophy, and engineering, Crozet stated that he would just teach engineering, applying only "the methods of instruction and authors of Ecole Polytechnique, especially those lessons which [he had] employed."[8]

Despite Crozet's lack of English, he introduced the first course of formalized engineering instruction at West Point. Partridge's unstructured and insolent approach to implementing the cadet program of education prevented Crozet and the Academic Board from standardizing the engineering course. Partridge routinely exercised favoritism and ignored faculty recommendations. Only after Thayer solidified his authority as superintendent in 1817 did the engineering course begun by Crozet expand into the full program of engineering education that became the centerpiece of West Point under Thayer. Crozet's tenure was not without its own controversies and friction. He resented Thayer's treatment of the professors, deeming Thayer's behavior "impolite." The French professor also complained of the geographic remoteness of West Point, and perhaps felt isolated as an outsider in the community of his American faculty peers. When Crozet openly challenged Thayer's conduct of cadet examinations, allegations of insubordination followed, leading to a short series of congressional inquiries. While there was an inconclusive court-martial, the reality was that Thayer guided the Military Academy with an iron hand to meet his vision of what the West Point program should be. Crozet departed in 1823 to become the newly appointed state engineer in Virginia.[9] With Crozet's introduction of engineering and Thayer's overall reform by decree, the Thayer system, as it became known at West Point, was thoroughly rooted in the precedent of the Ecole Polytechnique.[10]

The Thayer system was based on three tenets. First came academics, consisting of the scientific disciplines learned through a method of daily recitations by the cadets. Second was the standard of discipline to build character development. Third was the cadets' military training, done mostly with marching on the plain at cadet parades and during the summer encampment.[11] Under Thayer's system, every aspect of a cadet's performance in and out of the classroom was evaluated and ranked as objectively as possible, eliminating any hint of favoritism or political connection. At the end of each term, the academic rank of each cadet, by class, determined where and how he would go for the following year. Ultimately, this merit system was the final arbiter of which branch of the army the cadet would enter upon graduation and commissioning.

THE U.S. MILITARY ACADEMY ENGINEERING CURRICULUM UNDER THAYER

The two foundations of Thayer's academic program were mathematics and French. To adapt the Ecole Polytechnique system, West Point had to teach the cadets both subjects. With a solid understanding of math, the cadets could study engineering. The cadets had to learn French to read the texts purchased by Thayer and his predecessors. In the cadets' initial or fourth-class year, they studied math, which included algebra, geometry, and trigonometry, and French. Drawing and analytical and descriptive geometry were added in the third-class (sophomore) year. During the second-class (junior) year, cadets took drawing, which focused mainly on topography, as well as chemistry and philosophy. The second-year philosophy course included physics, known as "mechanicks," and astronomy. First-year (senior) cadets studied civil engineering and military art, geography, "tacticks," and chemistry.[12] Below is an example of what the West Point curriculum looked like during Thayer's tenure, in this case 1824 (see table). Note that the table lists instructors as well as textbooks used for each class.

During his time as superintendent, Thayer's design for the curriculum matured to include more specialized courses on civil engineering. In the 1820s, Thayer's Military Academy reluctantly added civil engineering to the curriculum to meet the calls for "internal improvements" required by America's expansion across North America. Thayer wanted the school to use engineering as a means for making competent military leaders, rather than producing officers who could also serve as engineers. Initially, Thayer and the Academic Board resisted Congress's demand that the Academy's engineering curriculum be augmented by a robust study of civil engineering in order to meet "the growing interest in internal improvements" in Washington, DC, and across the nation. As the nation and West Point sought greater purpose and validation for the existence of a military academy, the Board of Visitors suggested other benefits of having a nationally funded academy that produced educated civil engineers. After the superintendent, the Academic Board, which consisted of internally appointed department heads, was the main power controlling the scope and

Class	Department	Section	Name of Instructor	Course Names
1st	Fortification, Mil. Art, & Civil Engineering.	1st 2nd	Prof Douglas. Ast. Prof. Courtenay.	Gay De Vernon's Science of War, Fortifications, Sganzins Treatise on Civil Engineering, Perspective Shades and Shadows.
	Geography, History, Ethicks, National Law.	1st 2nd	Prof Picton " "	Morse's Geography, Tytler's History, Paley's Moral Philosophy, Vattel's Law of Nations.
	Tackticks.	1st 2nd	Maj. Worth " "	Rules and Reguilations for the Field Exercise and Manoeuvres of Infantry, Lallemand's Treatise on Artillery.
	Chemostry and Mineralogy.	1st 2nd	Doctor Percival	Cleveland's Mineralogy.
2nd	Philosophy.	1st	Prof. Mansfield.	Gregory's Mechanicks, Newton's Principia, Enfield's Philosophy and Astronomy.
		2nd	Lieut. Smith.	Bridges' Mechanicks, Enfield's Phil'y and Astron'y.
		3rd	Lieut. Mordecai.	Hutton & Enfield's Mechanicks, Enfield's Philosophy and Astronomy, and Haily's Philosophy.
	Chemistry.	1st 2nd 3rd	Lieut. Prescott. " " Cadet J. W. A. Smith	Henry's Chemistry.
	Drawing.	Whole Class	Mr. Gimbrede, Teacher. Cadet Catlin, Ast. Teacher.	Landscape and Topography. " "

Class	Department	Section	Name of Instructor	Course Names
3rd	Mathematicks.	1st	Prof. Davis.	Surveying, Descriptive Geometry, Perspective Shades and Shadows, Conic Sections, Biot's Analytical Geometry, Lecroix's Fluxions.
		2nd	Lieut. Webster.	Surveying, Descriptive Geometry, Perspective Shades and Shadows, Conic Sections, Biot's Analytical Geometry, Lecroix's Fluxions.
		3rd	Lieut. Green.	Surveying, Descriptive Geometry, Perspective Shades and Shadows, Conic Sections, Biot's Analytical Geometry, Lecroix's Fluxions.
	French.	1st	Mr. Berard.	Wonostrocht's Grammar,
		2nd	" "	Berard's Lecons Francais,
		3rd	" Du Commun	Gil Bias, Charles 12th.
		4th		
	Drawing.	1st	Mr. Gimbrede, Teacher.	Human Figure.
		2nd	Cadet Mackay, Ast. Teacher.	
4th	Mathematicks.	1st	Cadet Mahan.	Legendre's Geometry, Lecroix's Analytical Plane and Spherical Trigonometry, Descriptive Geometry.
		2nd	" Parrott.	Legendre's Geometry, Lecroix's Analytical Plane and Spherical Trigonometry, Descriptive Geometry.
		3rd	" Bache.	Legendre's Geometry, Lecroix's Analytical Plane and Spherical Trigonometry, Descriptive Geometry.

Class	Department	Section	Name of Instructor	Course Names
4th	Mathematicks.	4th	" McMartin	Legendre's Geometry, Lecroix's Algebra.
		5th	" McGehee	Legendre's Geometry, Lecroix's Algebra.
		1st	Mr. Berard.	Vonostrocht's Grammar,
		2nd	" "	Berard's Lecons Francais,
	French.	3rd	" DuCommun	1st Vol. of Gil Blas.
		4th	" "	
		5th	Cadet Findlay	

Source: Board of Visitors, *Annual Report of the Board of Visitors to the United States Military Academy* (1824), 16.

direction of the school. Thayer formalized the "academical staff" into a regulated Academic Board with substantial power and influence on the curriculum, examinations, merit, and awarding of degrees at West Point.[13] Unlike the Academic Board, the Board of Visitors was a federally appointed committee to provide civilian oversight of the U.S. Military Academy. On the Board of Visitors in 1821, Rufus King of New York argued that adding civil engineering to the mathematically focused curriculum would yield "greater public benefits" such as "constructing canals, roads, & bridges" and ensure that this cadre of officers would be employed in times of peace as well as war.[14] The Board of Visitors served as a bridge between the government and the school, and as such helped in the effort to expand the civil engineering curriculum.

The General Survey Bill passed by Congress in 1824 further emphasized civil engineering in the cadet curriculum. According to the bill, the president was authorized "to cause the necessary surveys, plans and estimates to be made of the routes of such roads and canals as he may deem of national importance in a commercial or military point of view, or necessary to the transportation of the public mail." Furthermore, the act empowered the president "to employ two or more skillful civil engineers, and as such officers of the Corps of Engineers" to complete the projects as he saw fit.[15] At the same time, railroad boosters sought to capitalize on West Point's emerging cadre of civil engineers who had entered army service. Railroad presidents deliberately promoted their projects to Congress and the president as having military value for moving men and supplies around the nation in the name of defense. Secretary of War James Barbour agreed to support Philip E. Thomas's Baltimore and Ohio Railroad construction with army engineers in 1827. Army engineers provided aid to at least twenty railroad companies during the 1830s.[16] In his annual report of 1828, Secretary of War Peter B. Porter remarked that "the Military Academy . . . has conquered all prejudices which formerly existed against it, and is scattering the fruits of its science, and communicating, by its examples, the lessons of industry and order there taught, not to the rest of the army, but to the youths of our country generally."

Moreover, Porter concluded that the Military Academy "will soon furnish every part of the country with the most accomplished professors in every branch of civil engineering."[17] He recognized that West Point's influence would ultimately increase the national capacity for the study and application of engineering. Over the course of Thayer's time as superintendent, the national government had driven the mission of the Military Academy toward providing an education much broader than one for purely military purposes.

However, it was questionable how much civil engineering and nonmartial coursework Thayer would allow Congress to force upon his curriculum. In 1825, Thayer explicitly opposed the addition of new courses to the school's curriculum. He wrote, "Those who are not satisfied with the existing course of studies have not reflected upon the nature and object of the Institution and have not considered that this is a special school designed solely for the purpose of a Military Education."[18] But Congress and several Academic Board members, particularly David B. Douglass, the first head of the Department of Civil Engineering, continued to link the future of West Point with the application of building internal improvements. Douglass embraced the idea that civil engineering was necessary for "civilian and military public works," and lobbied to travel to Europe and study civil engineering with the intention of bringing back a program and books modeled on "England, France, and Holland."[19] In the spirit of promoting the Military Academy, Thayer and the Academic Board came to accept the increased emphasis of civil engineering in the cadets' curriculum, but not in time for Douglass to be selected for travel to Europe. Instead, that opportunity was offered to Mahan, who would complete the incorporation of engineering as the central program of study for the cadets.

Although Thayer may have been hesitant to increase the amount of engineering in the cadets' studies, his leadership and example remained a permanent influence on West Point. As superintendent, Thayer was a model of discipline and efficiency. During his visit to West Point as a member of the Board of Visitors, George Ticknor recalled,

> Thayer is a wonderful man. In the course of the fortnight I have been here, he has every morning been in his office doing business from six to seven o'clock; from seven to eight he breakfasts; generally with company; then he goes to the examination-room, and for five complete hours never so much as rises from his chair. From one to three he has his dinner-party; from three to seven again unmoved from his chair, though he is neither stiff nor pretending about it. At seven he goes on parade; from half past seven to eight does business with the Cadets, and then from eight to nine, or even till eleven, he is liable to have meetings with the Academic staff. Yet, . . . he is always fresh, prompt, ready, and pleasant.

On that same visit, Ticknor noted about the effect of Thayer on the Corps of Cadets, "there is a thoroughness, promptness, and efficiency in the knowledge of the Cadets which I have never seen before, and which I did not expect."[20] Ticknor's study of Thayer captures the zeal and vigor that Thayer desired to impart upon the cadets. Colonel Thayer was the model graduate; not only was he on display for the cadets, faculty, and staff at the Military Academy, but he was also the paradigm of what kind of military professional the nation's military school was expected to produce. Thayer's self-discipline and adherence to order was replicated time and time again in the lives of the antebellum graduates, not just in uniform, but in their civilian professions as well.[21]

Reinforcing Thayer's personal example of officer professionalism was the Academy's famed merit roll. West Point's annual reports meticulously record the cadet ranking for every class and department throughout the antebellum era, beginning with Thayer's first year as superintendent.[22] After the highest-rated cadets were selected for a commission in the Corps of Engineers, the lower-rated cadets matriculated to the infantry, quartermaster, and field artillery upon graduation. While there were correlating trends between class rankings and how graduates fared as officers, success at West Point did not necessarily equate to later success in the army or elsewhere. For example, John

Newton finished second in the class of 1842 and was commissioned into the Corps of Engineers, while U. S. Grant, who finished twenty-first out of thirty-nine cadets in the class of 1843, received a commission in the quartermaster. While Newton went on to have success as an engineer in New York City, Grant's record is part of the American narrative.[23] However, the merit roll did ensure that the best performers in the scientific curriculum and in Thayer's system made it into the Corps of Engineers during the antebellum era. Additionally, the merit roll was a means to reward those cadets who displayed punctuality, self-discipline, and gentlemanly conduct. Finishing at the top of the class took on a reward and sense of entitlement in and of itself. The merit roll made sense in that it was a "scientific" way to select the best graduate of the nation's first "scientific" school. If daily recitation and blackboard exercises were the heart of the Thayer system, the merit roll was the logical evaluation of cadets in that system.

By the end of the examinations in 1826, Thayer's legacy to the Academy was firmly established. In the final remarks of their report for that year, the Board of Visitors observed: "In eighteen hundred and seventeen the system of instruction and discipline, now in practice, was introduced, by the present accomplished Superintendent, and has, by the Teachers and Academic Staff, been uniformly and consistently sustained; the favor of the nation has followed and encouraged their efforts; and now, every year, the privileges of this institution are sought for at the War Department, by above a thousand to whom it is not possible to grant them." Furthermore, the Board of Visitors argued that the U.S. "government should afford its Academic Staff a full and consistent support in their measures whether of instruction or of discipline."[24] In less than a decade, Thayer's system had gained the full support of Congress, the Board of Visitors, and most importantly, the Academic Board. Moreover, the demands for better civil engineering instruction drove the Military Academy to seek continued improvement in the quality of the civil engineering instruction. To that end, in 1827 the Academy sent Mahan, who at the time was a lieutenant and assistant to David Douglass, to Metz, France. As the next four decades proved, sending Mahan to France for a "graduate" experience in the study of civil engineering and the federal push for national improvements guaranteed that Thayer's Military Academy would have the best engineering program in the pre–Civil War United States.

Thayer was loyal to the nation and the Academy, but his sense of honor was paramount and explains why he departed West Point in 1833, never to return.

When President Jackson began to overturn disciplinary actions and expulsions that had been given by the Military Academy, Thayer thought that his authority as superintendent was being undermined. At one point, Thayer even complained that the president was overturning the Academy's disciplinary decisions for cadets who merely lobbied for favor in Washington, regardless of the infractions leading to a cadet's dismissal.[25] By most accounts, Jackson's open criticism of the school as an institution for the sons of privilege, and his repeated disapproval of Thayer's disciplinary decisions, contributed to Thayer's decision to depart West Point.[26] To the end of his life, Thayer was devoted to his sense of duty and his profession as a military officer. He demonstrated "unhesitating obedience to his superiors," even "yielding . . . his own opinion on subjects where his knowledge could be fairly challenged." Honor was the key to Thayer's character and judgment. Upon Thayer's death, George Ticknor recalled Thayer's character as being "good- tempered & gentle: clear-minded, far-seeing, always firm and, in matters of principle, unyielding; putting his country before everything else except his honor."[27] Thus, when Jacksonian politics and second-guessing reached the point of insulting his honor, Thayer could only resign and move on from his duties at the Military Academy. Regardless, by 1833 the foundation of the Military Academy's curriculum for the next half-century had been set.[28]

After 1833, Thayer continued his career as an army engineer in Boston, Massachusetts, and eventually retired to his home in Braintree, south of the city. Although Thayer never again set foot on the grounds of the Military Academy, he did remain in close correspondence with faculty who remained and many of the superintendents that came after. The father of the Military Academy was always looking to provide sage advice and recommend West Point graduates for academic posts at other new engineering schools as they were created.[29]

Late in life, Thayer continued to watch over the U.S. Military Academy from his home in Massachusetts. Paternal pride and affection marked much of his correspondence to George Cullum and others at West Point in the 1860s. Seeking to help repair any harm to West Point's post–Civil War reputation, Thayer made several recommendations, among them a proposal to form the Association of Graduates. Thayer envisioned this association as a permanent alumni organization dedicated to act as a "Board of Improvement" for the Academy, and in 1869, he was voted the first president of the association.[30]

Without Sylvanus Thayer, the ascension of civil engineering at the Military Academy and in nineteenth-century America would not have been as robust.

Thayer's example remained a touchstone for Military Academy faculty and graduates throughout the antebellum era.

MAHAN AND THE CIVIL ENGINEERING CURRICULUM

West Point's program of study remained largely unchanged after Thayer's departure because of Mahan's influence. Thomas Griess's 1968 dissertation, "Dennis Hart Mahan: West Point Professor and Advocate of Military Professionalism, 1830–1871," is still the best single comprehensive study of Mahan and his significance to the U.S. Military Academy. Griess convincingly shows how the cadet program of study in 1832 lasted through Mahan's tenure. The only exceptions were the curriculum adjustments made in the 1850s, when the Military Academy tested a five-year course of study in an attempt to include more military science and more engineering coursework. However, under Mahan's direction as professor of engineering and military science, the course content did change dramatically, especially as Mahan became more capable and adept at printing his own engineering textbooks and supplements. Griess's research finds Mahan to have been as vital to the success of the Academy as the Academy was to the arc of Mahan's life. Mahan taught the cadets "that precision of ideas, careful analysis, and hard work [were] essential to success."[31] And Mahan reinforced Thayer's emphasis on integrity and dedication to one's duty. The professor of engineering had the greatest influence on the cadets and West Point between Thayer's departure and 1861.

Dennis Hart Mahan's life in many ways represented the experience of Americans living in the era of the New Republic. His famous son, Alfred Thayer Mahan, recounted his father's story in his autobiography. The younger Mahan wrote that his father "was of pure Irish blood, his father and mother, already married, having emigrated together from the old country." The elder Mahan was a first-generation American, born in April 1802 shortly after his Irish parents arrived in New York City. In New York, he was baptized into the Roman Catholic faith at St. Peter's Church. When Mahan was a young child, his parents moved to Norfolk, Virginia, where Mahan grew up to become "a Virginian in attachment and preference."[32] From his son, we also know that Dennis Mahan had "begun the study of medicine ... in Richmond; but he had a very strong wish to learn drawing." This desire initially caused Mahan to seek admission to West Point in 1820. Dennis Mahan's admission into a military school was ironic in that he "drifted" to the civil side of teaching engineering and military art, caring not "for its pride, pomp, and circumstance."[33] The study of engineering as a means to produce military officers became Mahan's greatest passion and kept him at West Point for most of his life. From graduating at the top of his class in 1824 to heading the Engineering Department, Mahan found his identity and fulfillment at the Military Academy.

Mahan's approach to the study of engineering and the cadet program almost resembled the practice of a religion. He was a devout Christian, but even in his faith, Mahan was practical and, in a sense, rational. Mahan was unquestionably loyal, disciplined, and grounded in tradition, but he did not remain a Catholic in his faith. At some point prior to his commissioning in 1824, Mahan underwent a religious conversion, turning solely to the study of the Bible for his spiritual needs and guidance.[34] Writing to his stepmother, he encouraged her to tell his half-brother, Milo: "Above all things to stick to his [God's] book, tell no stories, & never speak bad of any person that by observing these rules every person will like him, everyone will believe him when he speaks and he may one day be a good man if not a great man, tell him that his brother has done everything for himself by his book (meaning the Bible) and that he must try and do the same."[35]

While this letter reveals insight into his strength of faith, there were certainly social and professional influences that also led Mahan to his devout Protestant rigor. More than likely this conversion was a result of both experiences at home and at West Point. His father's third wife, Esther, was a Protestant from Norfolk, Virginia, and the family became less drawn to the Catholicism of their native Ireland. Another influence was the mandatory sessions for cadets in the West Point chapel. At the chapel, the predominant denomination of military chaplains and army officers was Episcopal, although Presbyterian chaplains were a close second in preference for Academy leadership. It was no coincidence that Mahan's younger brother, Milo, ended up as a prominent Episcopal priest and professor of church history who taught at General Theological Seminary in New York City.[36] Throughout his life, Mahan lived and advocated strict self-discipline and control, as well as devotion to God and country. Mahan's devotion to the Episcopal faith provides a useful way to think about his passion for engineering in the West Point curriculum. Mahan's faith as well as his tenure as professor of engineering was disciplined, thorough, and practical.

In the course of his career, in addition to emphasizing engineering in the West Point curriculum, he also became a staunch advocate of the study of military history. Many of his students recalled how Mahan looked at history as a science that could produce and inform problem solving, just like any other science or mathematics. For example, Henry Halleck recalled learning as a cadet that "[it] is in military history that we are to look for the source of all military science."[37] Mahan wrote in his well-known military text, *An Elementary Treatise on Advanced Guard, Out Post and Detachment Service of Troops:* "Let no man be so rash as to suppose that, in donning a general's uniform, he is forthwith competent to perform a general's function; as reasonably might he assume that in putting on the robes of a judge he was ready to decide any point of law."[38] For Mahan, military science was just another scientific discipline to be learned and practiced in order to achieve an acceptable level of competency. Like other military theorists of the nineteenth century, Mahan believed that the "principles of war" studied in the science of war "transcended time and space" and applied to ancient civilizations as well as the United States.[39] After 1832, Mahan's course for the first class was "Military and Civil Engineering and the Science of War," the "capstone of the cadet's academic career."[40] In practice, however, the study of war and the military art remained a third priority behind civil engineering and the construction of field fortifications.[41] Mahan himself defended this prioritization in one of his texts, arguing that "[the] Military Art, in all its branches, is founded upon a comprehensive and thorough knowledge of the exact and physical sciences; and in no one branch is the importance of this knowledge more felt than in Engineering."[42] In Mahan's mind, the discipline of the study of engineering went hand in hand with the discipline of leading men in war.

Many cadets remembered Mahan for his repeated talks on common sense and the importance of being practical. Mahan evidently had a slight speech defect that led the cadets to sometimes refer to him as "Old Cobbon Sense."[43] In a letter to Frederick Harris, the father of David Bullock Harris, class of 1833, Mahan opined on how Academy graduates should carry themselves:

> There are some points common to most youth which I thought fit to caution him against, particularly those which seemed peculiar to young men from this school,—The first is, to forget that he has sprung full-fledged from a nest of Phenines, and to act, in consequence, like ordinary people, To court society, because he will there find the greatest amount and variety of that floating capital-ideas common to the mass called common sense, . . . [a] great fault of the graduates, they think they must look wisdom, talk

wisdom and forget, that the essential is to think wisdom and act like your fellow mortals.[44]

This concern with common sense manifested not only in cadet recollections of Mahan and correspondence; it also could be seen in Mahan's overall approach to the study of engineering in the context of educating future military leaders. Mahan advocated the firm foundation of an engineering education for all young men, and particularly when "common sense" and good judgment were lacking.

By the 1840s, Mahan's engineering course had matured to the point where observers and Mahan himself saw the low priority given to military science in the senior year course. The bottom line was that there was not enough time in the four-year curriculum to give adequate attention to civil engineering, fortification engineering, *and* military science. Mahan's philosophy was that the West Point cadet should master a few subjects well instead of studying many subjects with minimal comprehension. In his words, the cadets' curriculum "must be restricted to but a few subjects, that the mind may act healthfully and be developed by their study in its proper sense, and not merely crammed." Throughout his career he struggled with adding new content to the course. Given the choice of either being superficial or omitting the new material, Mahan usually chose to omit the new material. In the 1840 report of the Academic Board, Mahan detailed the components of his engineering program. He integrated carpentry, stonecutting, and basic mechanics into the program. Cadets also took sub-courses on construction materials, foundations, and arches. Mahan's robust course also focused on engineering theory, and all of this material was to be mastered in the cadet's senior year.

Mahan was exact and taxing with his students, especially in drawing. He demanded that the cadets produce precise schematics and drawings for each of the construction studies, ranging from timber to rock. All cadets first took drawing during their second year, and each year thereafter. According to Mahan, mastering the pen, pencil, and ruler were the rudimentary skills to successful engineering. An officer needed to be able to draw a design that any builder could use. Since architecture was still a nascent field, engineers were expected to draft all the designs for any given project. All army officers needed to be able to draft fortification plans or copy other plans efficiently. Additionally, West Point instructors could use drawing tasks to evaluate cadets at the boards during recitations.[45]

There were practical military reasons for having cadets take drawing as well. Prior to the advent of photography in the 1840s, drawing was the only way to portray the battlefield in a sketch. When building a fort, a detailed sketch had to be made before construction could begin. Rendering an accurate sketch was part of how a good military leader communicated, and thus, drawing was an indispensable course in the West Point curriculum.[46]

West Point's drawing curriculum became firmly fixed with the appointment of Robert Walter Weir to the Military Academy in 1834. For the first three decades of the Academy, the drawing course changed from year to year depending on who taught it. Weir, at the age of thirty-one, was a well-known artist in New York who had traveled to Italy and was well connected in New York's social circles. The romantic poet William Cullen Bryant recommended Weir to Secretary of War Lewis Cass in 1834, writing, "Mr. Weir possesses a high reputation among his brethren of the art . . . in the several departments of portrait, landscape, and historical painting."[47] Based upon this recommendation and Weir's reputation, the War Department and President Jackson hired Weir. Weir would hold the position for forty-two years. Just as Mahan came to represent the Department of Engineering, Robert Weir came to exemplify the drawing course and its institutional foundation. According to George Cullum, "Weir's methodical habits, devotion to duty, elevated character, dignified bearing and eminent professional reputation, soon established him in the hearts of officers and cadets, who regarded him with the greatest pride and sincerest affection."[48]

Under Weir's guidance, the drawing course transformed from the common art school practices of sketching with crayon and pencil to a more disciplined course in mechanical and engineering drawing. Although topographical drawing remained a key task, the cadets also had to master the more technical aspects of drafting and design. Weir improved the quality of drawing instruction by adding plaster copies of ancient statues and constructing a larger room where cadets could create and show their artwork. While Weir understood the value of traditional drawing instruction, he also made extensive use of basic art and architecture texts. The most influential textbook he used was Seth Eastman's *Treatise on Topographical Drawing* (1837). Eastman, class of 1829, had studied under Thomas Gimbrede, one of Weir's predecessors, and became Weir's assistant from 1834 to 1840. A commissioned army officer, Eastman understood the importance of drawing, topography, and engineering for the cadets. He developed a standardized system of "symbols, shapes, and shadings" that became the standard not only for West Point but for the rest of the army.

Eastman most likely based his ideas on "outdoor sketching trips" he took with Weir. Eastman's text and map legends were the core of West Point's drawing program in the nineteenth century.[49] Eastman's system of topographical drawing became the standard for American topographical engineers as well.[50]

In addition to extensive use of Eastman's book, the graduates learned and practiced the standardized topographical symbols and drawing. Based upon Morrison's calculations, cadets spent almost seven hundred hours drawing during their time at West Point. Second- and third-class cadets each spent 250 hours drawing, and first-class cadets spent 192 hours drawing in their courses on civil engineering and military fortifications.[51] Regardless of what branch of the army the cadet entered, he was sure to be well trained in topographic drawing and sketching engineering designs.

Over the course of their overlapping careers, Mahan and Weir became more than professional acquaintances. Mahan complained to Academy leaders that his and Weir's pay were both below that of their peers teaching mathematics and science. Consequently, the school promoted Weir to professor of drawing with a corresponding pay increase. During this time, Weir and Mahan became friends while supporting each other's courses in teaching the curriculum.[52] No doubt Mahan's early desire to learn drawing, a desire that brought him to West Point as a cadet, helped strengthen the relationship between the two men. Their common stands in administrative matters and in curriculum development made architectural drawing an integral part of the engineering program at West Point.

The U.S. Military Academy provided the "first architectural education for engineers" in America with the start of the engineering course as taught by Crozet. As the course matured under Crozet and David Douglass, the architectural aspects of the engineering course became the most developed under Mahan. Mahan included French architectural theory from Jean Rondelet and Jean-Nicolas Durand, as well as Quatremere de Quincy's theory of character, into one "comprehensive curriculum." Mahan's "emphasis on functionality, practicality, and appropriateness" shows a direct influence from the French masters and was incorporated into Mahan's textbooks, which can be considered America's first architectural textbooks. He taught the West Point engineering course as "the most advanced theoretical background of any civil engineer in America."

Marvin Anderson's recent analysis of Mahan's textbooks shows an evolution in Mahan's understanding and use of architectural theory in civil engineering.

When he began teaching in 1831, he used an old elementary French engineering book by Joseph Mathieu Sganzin and his own 70-page text, *Supplements to Sganzin*, which centered on "bridges, dams, and canals." In less than three years, Mahan wrote his own 172-page textbook, *Outlines of the Course of Civil Engineering*, complete with supplements and a 16-page *Notes on Architecture*. By 1837, Mahan's texts had grown into a popular engineering text, *An Elementary Course of Civil Engineering*. He wrote this text after combing through various European and American engineering textbooks and making numerous visits to construction sites in early America. Throughout all of Mahan's books, he emphasizes the practical and functional features of various architectural styles, preferring the Greek to the Roman, and condemning the Gothic style overall. Visitors to the West Point campus today might find this treatment a bit ironic, as the current Mahan Hall is built in the neo-Gothic style.

Anderson concludes his examination of Mahan's approach to architecture by noting that the West Point professor "focused on principles and materials as well as theories and methods of architectural design: he presented architecture as an integral part of the engineering discipline." At the end of his career, Mahan devoted more class time to architecture, again emphasizing the practical aspects of architecture through the buildings recently constructed by the Corps of Engineers at West Point.[53]

As a teacher, Mahan followed the tradition of the "Thayer Method" in that he supervised the cadets' daily recitations and rarely lectured to the class. Morning sessions with sections of twelve cadets generally lasted ninety minutes each, but afternoon sessions with the entire class doing engineering drawing could last three hours. If Mahan was not teaching a class of his own, he would visit the classes being led by his assistant professors and formulate additional questions and comments for the class.[54] In the first-year (senior-year) course on engineering and science, Mahan's textbooks made up the majority of the dense reading load, so whatever influence he failed to have in the classroom, he made up for with reading assignments steeped in civil engineering and fortification principles.[55] In the classroom and in his writings, Mahan followed Thayer's example of being "stern and unyielding where duty was concerned." No principle could be compromised, and properly attained intellectual growth also cultivated "qualities of discipline, integrity, loyalty, and honor."[56]

Mahan's basic philosophy of study "stressed the vital importance of self-reliance and thorough grounding in basic principles." In a letter to Cadet George Welker, he wrote,

I should recommend, however, but one course, which is not to burden your-self with too many books, nor to rely upon any authority without first put-ting the reasoning to the test of rigid examination—for this is the only means to strengthen the judgment and to acquire that most essential of good habits, self-dependence. The best plan, I think, that a young engineer can pursue is to read everything connected with his profession that falls under his eye, and to note every new useful idea under its proper head in his notebook, with the authority, not copying the text but the spirit.

With this approach, the talented cadet would be able to bridge theory with practical application. Unlike French engineering schools where students first studied and mastered theory prior to progressing to a graduate school of appli-cation, West Point cadets had to accomplish both as an undergraduate (cadet).[57] Some would even say that Mahan's course overlapped into the discipline of mechanical engineering.[58] Again, Mahan's goal was to produce the best-trained engineer for the officer corps in the finite time allotted by the program of study. The skills and discipline of an engineer were thought to be transferable to any profession, and most importantly, the military profession. According to the West Point faculty, the cadet curriculum was a "strict course of mathematical and philosophical study, with applications to the various branches of military science."[59] A graduate trained in these basic principles could easily serve as an officer in the infantry or cavalry as well as in the engineers.[60]

As for Mahan's instruction on military science and the military art, Na-poleon was the central figure from history whom the cadets studied. Mahan's interpretations of the Napoleonic wars introduced American cadets to the French way of war in the nineteenth century, and created such an aura that even the military feats of Gen. George Washington were overlooked at the Academy. Despite its lack of strategic insight, Mahan's *An Elementary Trea-tise on Advanced-Guard, Out-Posts, and Detachment* proved to be an influential text in nineteenth-century American military thinking.[61] Similar to Carl von Clausewitz and Antoine-Henri Jomini, Mahan's military theory is marked by a thoughtful reflection on the warfare waged in Napoleonic France. Mahan's writings on military theory, limited as they were, connected to the tenets of scientific and mathematical thought. Remembering Mahan at the National Academy of Sciences, Henry L. Abbott (class of 1854), noted that "[in] the great game of war the field is more extended, and the skillful player must com-bine the precision of the mathematician with the profound knowledge of the strength and weakness of human nature. Nothing can be neglected."[62] Dennis Hart Mahan's use of engineering to make military leaders remained a constant in his program of study for the Corps of Cadets.

After the Civil War, the Board of Visitors became concerned about Mahan's health and his effectiveness as professor of engineering at the Academy. In its June 1871 report, the Board of Visitors recommended that Mahan be retired, even though President Grant had let Mahan remain past the age of military retirement at sixty-nine. By September 1871, prone to depression and melancholy, Mahan committed suicide by jumping into the paddlewheel of a Hudson steamboat carrying him to New York.[63] Those who memorialized Mahan reflected on his influence on the cadets he educated, his contributions to engineering in the United States, and his service to the U.S. Military Academy.[64] In a sense, Mahan's drowning coincided with the end of West Point as the leading higher institution of engineering education in the United States.

Thomas Griess concludes that "through the military academy, Professor Mahan made his greatest contribution to military professionalism." Griess also acknowledges the "hundreds of engineer officers" who "learned well the rudiments of their profession" from Mahan.[65] Mahan's contribution to engineering in antebellum America is in many ways a more enduring legacy.[66] When taken as a whole, the textbooks, the postgraduate careers of his students, and over four decades of engineering education at West Point all affected the growth and expansion of the engineering field in the United States. *An Elementary Course of Civil Engineering* went through several editions and became the standard handbook for American engineers in the nineteenth century. Over fifteen thousand copies are estimated to have been printed and circulated throughout the engineering institutions of the United States and around the world by the time of Mahan's death.[67] Twenty percent of the 1,887 Academy graduates between 1802 and 1860 went into civilian engineering at some point after graduation.[68] Between 1820 and 1860, civil engineers in the United States transitioned from all being self-taught to more than half graduating from engineering schools influenced by West Point.[69]

The accomplishments of the Military Academy and its antebellum graduates point to Mahan's, and thus, West Point's influence on engineering. The breadth and number of institutions and professional engineering organizations that were touched by at least one West Point graduate suggest a unique influence across antebellum America. Men from West Point headed engineering programs at Harvard, Yale, Columbia University, and the University of Michigan.[70] By 1840, forty-nine graduates "had been appointed chief or resident engineers on railroad or canal projects." Eleven of the original fifty-five members of the American Society of Civil Engineers formed in 1852 had graduated from West Point.[71] Thirty-five West Point graduates went on to become president of a college or university in the nineteenth century.[72] Given this legacy, Henry Abbott's memorial is less hyperbole and closer to reality. Abbott wrote that Mahan was "regarded as one of our leading pioneers in scientific culture—one who has laid the foundations upon which many have founded titles to lasting fame."[73] Perhaps the greatest compliment to Mahan came from Sylvanus Thayer several years before Mahan died. Thayer wrote the president of Dartmouth College that Mahan was one "who stands highest among the distinguished West Point professors."[74]

AN ANTEBELLUM ENGINEER CORPS

Antebellum West Point sought to create an army officer corps that could remain obedient to the directives of a civilian, republican government. In order to ensure that the army would not become an institution of tyranny or corruption, the U.S. Military Academy embraced the utility and practicality of a scientific and mathematical curriculum. By giving the military student a means to function outside of fighting wars, the Academy and the United States ensured that the nation did not have to pay for an idle military force. Over time, the growing cadre of officer graduates proved that they could assist with internal improvements of the young nation. West Point and its civilian masters in the federal government embraced the engineering program that Thayer and Mahan had imported from the Ecole Polytechnique. West Point men were not only expected to defend and fight for the United States; they were also expected to literally build the nation.

Thayer created and standardized the Thayer system of discipline and recitation. Mahan made sure that the system would produce engineers. While the engineering success of an Academy graduate was not preordained upon completing West Point, the foundation in engineering enabled those who did succeed to do so immeasurably. Graduates of the Military Academy used various elements of engineering theory from Mahan's courses to build railroads, survey and map the topography of the United States, establish new schools for engineering, and contribute to the professionalization of the engineering field. West Point's engineering curriculum was the foundation for engineering education and improvement in the United States.

The purpose of the U.S. Military Academy led to its emphasis on science, math, and eventually engineering, but that purpose also restricted how far the institution could evolve in its scope and curriculum. Dedicated first and foremost to producing military leaders to fight and defend America in war, the school always had to balance that mission with its teaching method. Thayer resisted the push for more engineering coursework for fear of losing that military focus. Mahan and the Academic Board struggled to have the proper balance between military science and civil engineering. The school even experimented with a five-year curriculum to achieve all ends and desires for a well-developed soldier-engineer. Antebellum America validated West Point's curriculum in that so many graduates contributed to the country's internal improvements as it expanded westward. Moreover, the Mexican-American War proved that West Point graduates could defend the country when called upon.

The Civil War was more problematic for the Military Academy. On the one hand, the majority of the school's graduates ably reentered military service and led the Union to victory. On the other hand, graduates who defected to the South opened the Military Academy to legitimate criticism of its purpose and effect. The institution that had created a body of engineers and military officers had failed to create unanimous loyalty to the nation. After the Civil War, the nation and the Academy realized that there were limits and risks to the program of study. As a result, West Point and it supporters emphasized the accomplishments and sacrifices of the Union graduates in the three decades after the war. Discipline waned and the curriculum stagnated, leaving newer, emerging universities to take the lead in engineering education in the United States.[75]

The antebellum legacy of West Point's engineering program, however, endured through the end of the nineteenth century. Men who had studied under Thayer, Mahan, and the West Point faculty continued to build railroads, roads, and parks after the war. As veterans of the Civil War, they served in Congress, formed veterans' groups, and helped establish professional organizations. Arguably the Civil War and their postgraduate experiences had a greater effect on the lives of the West Point men, but it remains that their education as cadets was common to all. The cadet experience remained a benchmark for these men, and indirectly, for any endeavor they pursued. In that light, perhaps the Board of Visitors of 1867 was right to claim that "the United States Military Academy is not an institution for the benefit of a favored few; nor should it be an experimental arena of the youth of our country. It belongs to the nation, and is supported for the nation's welfare."[76] And that support would become acutely manifest in New York.

4

ENGINEERING
PROFESSIONALS IN NEW YORK
A New American Identity

Moreover, the Croton Water is slowly flowing towards the city, which at last will stand a chance of being cleaned—if water can clean it . . .

—GEORGE TEMPLETON STRONG,
28 June 1842, from *The Diary of George Templeton Strong*, Vol. 1

In late October 1825, Governor De Witt Clinton of New York made a historic journey on the Erie Canal, traveling by barge from Buffalo to Albany and then down the Hudson River to Manhattan. At the end of his ten-day journey, Clinton symbolically "wedded the waters" by pouring two casks of water from Lake Erie into the Atlantic Ocean.[1] When Clinton sailed past Thayer's Military Academy at West Point that first week of November, no one could have foreseen the role that Thayer's cadets would have in exploiting the commercial boom in New York created by the Erie Canal. In less than three years, freight flowed over the waters of the Hudson and the Erie Canal at twice the rate of cargo flowing on the Mississippi through New Orleans.[2] By the mid-1840s, New York's cheap and efficient transportation system made Gotham the commercial capital of the United States, outpacing Boston, Philadelphia, and Baltimore.[3] New York's continued commercial success depended on the city's ability to accommodate the rising class of a moneyed elite, middle-class professionals, and a working class with an improving quality of life in New York. Encouraged by the wealth of the canal boom, city leaders pursued an aggressive program of building and development that drew West Point graduates to the antebellum metropolis. The Erie Canal, or "Clinton's ditch," not only expedited the economic growth of New York City, but also was a catalyst for urban infrastructure projects and engineering.

Of course there were other factors beyond the Erie Canal that contributed to the dramatic rise of New York in the nineteenth century. The city benefited from a decision by British merchants after the War of 1812 to dump three years' worth of exports that had been sitting idle in British docks in New York City rather than in Boston or Philadelphia. New York merchants responded by enacting a very liberal auction law, which mandated the sale of many cargoes by auction—no matter how low the bid. This perceptive move enticed hundreds of southern and western merchants annually to New York City after 1815, and many entered into contractual shipping and credit agreements with Manhattan (not Boston or Philadelphia) merchants. Other innovations contributed to New York's commercial fortune. In 1845, Cincinnati completed the Miami and Erie Canal up to Lake Erie, effectively connecting the Ohio and Mississippi Rivers to the Erie Canal and New York—giving the city an all-water route to New Orleans (whose importance was greatly enhanced after the advent of Robert Fulton's steamboat).[4] Early American diplomacy also underwrote the rise of New York's commercial triumph. Both the opening up of the Mississippi–Caribbean Sea route in 1795 following the Pinckney Treaty, and the U.S. acquisition of Florida via the Adams-Otis Treaty in 1819, removed Spanish vessels from Gulf coast waters and the Florida Keys, making it easier and safer for ships bearing cotton, hemp, rice, indigo, sugar, and tobacco to sail from New Orleans to New York. This safer transit north transformed New York City into America's garment center and a major producer of cigars, cigarettes, fine candy, cakes, and many other goods, giving the city a vibrant commercial and industrial economy by the 1830s.[5]

And then there was the reality of geography and location that "predestined" New York's position. Rival Philadelphia (and the Pennsylvania Turnpike that it and Pittsburgh helped build) had to climb the Allegheny Mountains west of Pittsburgh with a less efficient inclined railroad. For Baltimore, the National Road provided an initial advantage, but that route also had to negotiate the Appalachian Mountains, and by horse and carriage. Again, New York possessed favorable geography, having the only low spot or notch in the mountains between Georgia and Canada, the terrain through which the Erie Canal ran.[6]

As the first major infrastructure project in the nation, the Erie Canal was the first important engineering contribution to the growth and expansion of the United States in the nineteenth century. Without this initial feat of connecting the port of New York City with the interior of the young republic, there might not have been as rapid a growth rate westward. The economic power generated by the canal, followed by the proliferation of railroad lines to the nation's core Atlantic cities, fueled New York's and the nation's development for over six decades. Commercial goods flowed east and west along the Hudson River, the Erie Canal, and the Great Lakes route, creating a vital line of economic growth between Chicago and New York.[7] Freight totals on the canal eclipsed railroad totals in New York through the 1850s. From 1825 to 1882, the Erie Canal generated revenues totaling more than $121 million, and supported twenty million people.[8] For nineteenth-century Americans, the canal, the railroads, and the subsequent development of New York City's infrastructure validated the premise that science and technology in the hands of the engineering profession could fulfill the economic potential of the country.

Such economic advances came with a price. They led the New York legislature to use heavy-handed supervision to preserve a sense of civil order, build an effective sanitation infrastructure, promote health, and prevent fires.[9] A lack of clean water led to the cholera outbreaks and uncontrolled fires that disrupted the market forces expanding New York. City leaders looked to the pure and abundant water of the Croton River, diverted to the city by engineers, to cleanse Manhattan and transform New York into a more ordered commercial capital of American and Atlantic trade.[10] In large part, the Croton Aqueduct System enabled the Board of Aldermen to meet many of its safety and sanitation responsibilities. The city's political leaders needed a cadre of technically competent experts to build the aqueduct as well as a myriad of other public works programs such as Central Park, an integrated sewer system, paved streets, the Brooklyn Bridge, and similarly ambitious projects.

Besides the economic and political outcomes of the new infrastructure and engineering projects in America's urban areas, these projects also enabled the organization of the engineering profession in the United States. Building public works and infrastructure provided a focal point for the practitioners of science and engineering to come together, delineate their expertise, and identify qualifications for membership. Engineers sought a virtuous position in the new republic, one that would follow the promise of technology and the path to a transcendent society in the United States.

The first generation of West Pointers to come to New York staked their careers on their professionalism and reputations, seeking to remain apolitical and true to the principles of "science and engineering." Among the first West

Pointers answering the summons for engineers to New York were a professor and two of Thayer's students. David B. Douglass, professor in the Engineering Department at the Military Academy between 1816 and 1831, had supervised a portion of the Erie Canal construction and several other private canals.[11] Between 1833 and 1835, Douglass used his limited practical expertise, which he had gained through his canal experience, to become the chief engineer of the Croton Aqueduct.[12] William Sidell, class of 1833, worked under Douglass in 1834. George S. Greene, class of 1823, worked on the Croton Aqueduct before and after the Civil War.[13] Unlike the first generation of Academy graduates who went to New York City, the graduates who came in the 1850s and after the Civil War were more drawn to the political machinations of the city and the state. But the earlier West Point engineers, Douglass, Greene, and Sidell, nobly pursued engineering feats, using science to improve the lives of New Yorkers. In the process, they also managed, perhaps to their detriment, to remain above the fray of antebellum city politics. This idea espoused by the earlier graduates of the Military Academy, of practicing civil engineering for a common good, emerged within the urbanization and modernization taking place in New York and made the city the center of nineteenth-century enterprise, innovation, and identity for American civil engineering.

As the largest city in nineteenth-century America, New York naturally presented the most pressing challenges to civic order and municipal infrastructure. Besides getting Croton water to the city's masses, civil engineers responded to the boundless opportunities an expanding metropolis offered for road construction and infrastructure development, maintenance, and improvement. It would be no coincidence that seven of the first eight men to lead the American Society of Civil Engineers had worked as city employees. City leaders relied on the proficiency of the engineers to get things done. Regardless of whether they were West Point–educated or trained on the early canals or railroads, New York engineers forged an identity of expertise, reliability, and to a certain degree, partisan independence.[14]

During this same period, other professions also sought to organize and advance their expertise, but they were not as reliant on the allure of New York's projects for gaining support. For example, in 1871, mining professionals and businessmen in Wilkes-Barre, Pennsylvania, established the American Institute of Mining Engineers (AIME) for those engaged in any aspect of the mining industry. Led by the spirit of the Centennial Exhibit in Philadelphia, Alexander Holley and mechanical engineers formed the American Society of

Mechanical Engineers (ASME) in 1880.[15] The miners' organization and the mechanical engineers' society were reactions to the realities of post–Civil War capitalism, industry, and national pride. For civil engineers, though, it was the combination of the geographic focus and the desire to create an organized body of identifiable experts that enabled them to emerge as one of the first clearly recognizable professions in the United States.

A key element to the engineers' professional identity was their sense of place in American society. Burton Bledstein argues in *The Culture of Professionalism: The Middle Class and the Development of Higher Education in America* that professionalism "was a culture—a set of learned values and habitual responses—by which middle-class individuals shaped their emotional needs and measured their powers of intelligence."[16] Essential to Bledstein's understanding of professionalism was the rise of the American middle-class identity in the mid-Victorian era. Citing the observations of nineteenth-century Frenchmen Michel Chevalier and Alexis de Tocqueville, Bledstein argues that the concept of "middle class" was unique to antebellum American society. For many Americans, being middle-class connoted a standing in society that had the potential for material improvement as well as increased social stature. In nineteenth-century America, one's class no longer limited one's ambition or potential for upward mobility. If one could attain wealth and consumer goods, then one could appear to be better than the place from which he came. But merely attaining wealth and looking the part was not enough. One had to maintain his improved stature through the creation of and adherence to an "institutional order."[17] Through the process of creating that order, professions emerged in almost every calling in life. Men in the field of civil engineering were among the first to espouse institutions of professional association, practices, and expectations of professionalism.

Depending on the scope and view of the historical study, defining the middle class, especially in the nineteenth century, is tricky. The middle class can be identified by one's relationship to production in the economic sense, but it also can be associated with one's sense of "class consciousness." Jennifer Green offers a useful way to look at the meaning of "middle class" in her study of southern military academies. Green adapts E. P. Thompson's assertion that class is dependent upon one's own history. She describes "class in a particular time and place in the process of class formation. Indeed, the examination of class compels us to make a static group out of one constantly in flux." Thus, individuals may be counted in one or more groups based upon context. Green also applies

Max Weber's definition of social class, which is based on one's relationship to the means of production. But unlike Weber, who says the "owners" are the "middle class" and the laborers are the "working class," Green says that "class development is based upon economic and occupational structure." Moreover, in the United States, individuals did not identify so much with a social class consciousness in the nineteenth century as they already possessed a sense of "class awareness." Thus, the American middle class in the nineteenth century was a class identity based upon an awareness of one's economic status and vocation in relation to the social zeitgeist of the emerging United States.[18]

In his study of Jacksonian America, Charles Sellers describes how the term *middle class* came to name both a self-consciousness and a moral point of view. A good middle-class American was "hard-working" and practiced "self-discipline" on the way to becoming "self-made." Using economic criteria to define class, Sellers shows how the reality of being middle-class was a myth for most Americans in the Age of Jackson. When class was connected to the relative wealth of an individual, the richest 10 percent of Americans controlled 73 percent of the national wealth by 1860, with the largest increase of wealth concentration occurring between 1820 and 1860.[19] According to Stuart Blumin, in the United States there was a sense of being middle-class based upon one's annual income. Blumin quotes Walt Whitman's 1858 commentary: "The most valuable class in any community is the middle class, the men of moderate means, living at the rate of a thousand dollars a year or thereabouts." Yet Blumin's main argument is that the term *middle class* as such was a "misnomer" because the overarching struggle between "formerly aristocratic upper class and a decidedly plebian lower class" waned as American culture writ large came to embrace "bourgeois values."[20] Here, Sven Beckert's interpretation of class status and the city is most useful. Like Blumin, Beckert argues that *middle class* is overused, and the term that best defines the subject is *bourgeoisie*, which he "uses interchangeably with 'upper class' and 'economic elite.'" They were a "particular kind of elite whose power, in its most fundamental sense, derived from the ownership of capital rather than birthright, status, or kinship." More importantly for this study, Beckert's definition of "bourgeoisie" consists of two central groups, the "moneyed" elite and the "professionals." The first group Beckert classifies as those large "merchants, industrialists, . . . bankers" and "rentiers" or "people who lived off investments they did not manage themselves." Beckert's second group of the "city's bourgeois" were the "professionals, experts, and intellectuals." In New York, the West Point men decidedly were in this second group when they first entered the city's society, but like many of the professionals Beckert examines, they too complicated the relationship by finding "access to bourgeois networks and bourgeois institutions solely based on the educational capital they controlled." Thus, the military men who made the transition to the city possessed the socioeconomic traits of the "middle class," and they had the capacity to ascend to the elite class of "*entrepreneurial bourgeoisie*" in nineteenth-century New York.[21]

Finally, with regard to the American middle class, Robert Wiebe's observations are helpful in understanding the connection of nineteenth-century professionals and the rise of the Progressives. Professionals in urban areas wrestled with the "confusion" created by a modernizing world, and then looked to "science" and a "romantic" sense of "nature" to resolve the challenges created by modern America. These professionals, Wiebe writes, formed a new class, a middle class, that would eventually come to embrace Theodore Roosevelt.[22]

For the purposes of the discussion of the cadets' socioeconomic class, "middle class" generally means the population group that is between the laboring class and the wealthy elite. According to this general definition, the majority of men who came to West Point were arguably middle class. James Morrison's study of the Military Academy from 1833 to 1866 used the school's own criteria to approximate the backgrounds of the cadets. Morrison notes that over two-thirds of the cadets' parents were farmers, merchants, lawyers, army officers, and planters. A key attribute of these occupations is that the individuals controlled the means and ends of their work, as opposed to a day laborer in a shop or factory. While farmers were the largest concentration of West Point parents at 24.8 percent, Morrison comments that the percentage was still smaller than the farmers' share of national population, which was 44 percent. Morrison also found that other than some "Lees, Herberts, Du Ponts, McAlesters, and even a Bonaparte," elite American families did not send their sons to the Military Academy.[23] To be sure, the sons who sought and received appointments to West Point were seeking to maintain or improve their place in society. In this sense, the cadets represent Bledstein's middle-class criteria in that they "acquired ability, social prestige, and a life style approach [to their] aspirations."[24]

In order to refute the Jacksonian accusations of elitism and privilege, West Point tracked the "Circumstances of Parents of Cadets," classifying them as being "Indigent, reduced, Moderate, or Affluent." The "moderate" cadets fit the accepted definitions of middle-class. Between 1842 and 1879, 83 percent of West Point's graduates came from "moderate" families, with a smaller percentage coming from rural areas, defined as farms, than the overall national percentages.[25] The young men arriving at West Point were an educated group coming from farms, towns, and cities. After they decided to pursue an Academy appointment and become cadets, these men embraced the standards and code of behavior demanded by the Military Academy as a means to graduate, become officers, and attain the status of a respected military professional. Granted, there were other reasons to seek an appointment to the United States Military Academy. Cadet candidates often cited economic hardship and the promise of free education as the reason for seeking admission. Others sought "martial glory" and political influence. Some saw prestige in receiving the appointment from their sponsoring congressman. Regardless of the motive for pursuing an appointment to West Point, the Military Academy became "a powerful institution of professional socialization."[26] As explained in chapter 3, the West Point program and science-oriented curriculum predisposed the graduates to order, discipline, and engineering.

For the graduates in antebellum New York, conforming to the standards and expectations of professional engineering was a familiar way of life. Bledstein asserts, "[in] the years before the Civil War, an aspiring middle class in America was beginning to build a professional foundation for an institutional order, a foundation in universal, scientific, and predictable principles."[27] In light of the engineering program led by Thayer and then Dennis Hart Mahan, the professionalism movement made sense to the West Point men. West Point taught cadets to make the unpredictable nature of war predictable through the deliberate application of scientific thought and "military science."[28] When Bledstein writes that "[the] professional penetrated beyond the rich confusion of ordinary experience, as he isolated and controlled the factors, hidden to the untrained eye, which made an elaborate system workable or impracticable, successful or unattainable," he could easily have been paraphrasing Mahan, or Jonathan Williams, the first superintendent.[29] Even though the primary identity of a West Point graduate was that of an army officer, the graduate remained a product of his socioeconomic background. As a civil engineer and a military officer, the West Point graduate merged the two identities into a sense of professionalism for engineers in mid-Victorian America. If an Academy graduate could achieve one type of status and respect as an officer and alumnus of West Point, then he might attain even greater prestige and status by propagating his engineering skills with his civilian peers. The title of "civil engineer" could further distinguish him from his moderate or middle-class origins. Moreover, belonging to a select group of individuals trained in engineering replicated the sense of camaraderie and importance each graduate experienced as an army officer. Finally, being professional meant these men were "unambiguously at the center of New York's bourgeoisie ... where high-ranking civil servants, military officers, church officials, and state-employed professors often constituted an important part of this class."[30]

FIRST ATTEMPTS TO PROFESSIONALIZE
AMERICAN CIVIL ENGINEERING

The American Society of Civil Engineers officially began at the Croton Aqueduct offices in Manhattan's Rotunda Park in late 1852.[31] Prior to that year, proponents of professionalization in engineering were dispersed around the United States. Although there is some evidence showing that the "earliest effort to form an association of the civil engineers" may have started in Augusta, Georgia, the first effort of record occurred in Baltimore, Maryland, at Barnum's Hotel on 11 February 1839. Invited by the Maryland Academy of Science and Literature, forty men from different states elected Benjamin H. Latrobe of Baltimore as president of the convention.[32]

Latrobe was the chief engineer of the Baltimore and Ohio (B&O) Railroad.[33] The B&O was the first major improvement project to attract a substantial number of engineers. It was also Baltimore's opportunity to outflank New York in accessing the markets in the Midwest, an endeavor that would fall short because Baltimore could never really influence and control the markets and business in the West the way New York did. In building a railroad from Baltimore to Ohio, following the Potomac River, the B&O also leveraged the provisions of the Survey Act of 1824. When the B&O started construction in 1828, the company sent engineers to England to study railroads already in operation. After hiring Jonathan Knight and Caspar Willis Wever to be the "engineer and superintendent of construction respectively," the railroad board sought a team of qualified topographical engineers. Eight of the ten topographical engineers were West Point graduates, while the other two attended the Academy but did not complete their studies.[34] All eight of the West Point graduates were still active-duty army officers when the B&O Railroad hired them in 1828. This use of army officers was in accordance with a Survey Act provision that authorized the president "to employ two or more skilful civil engineers, and such officers of the Corps of Engineers, or who may be detailed to do duty with that corps, as he may think proper."[35] The B&O Railroad was the first company to request army engineers from the War Department and to receive them from the army.[36] Consequently, by 1839, when Latrobe and his fellow civilian-trained engineers decided to create a professional order of engineers in Baltimore, they had already shared a decade of engineering work with the West Point–trained men.

From the Baltimore convention, a committee of seventeen met in March 1839 at the Franklin Institute in Philadelphia to write a constitution that would govern the association of civil engineers. On the committee were seven men associated with the U.S. Military Academy. William Gibb McNeil graduated in 1817. George Washington Whistler, father of the famous romantic painter, graduated from West Point in 1819. Benjamin Wright, Walter Gwynn, and Isaac Trimble were members of the class of 1822 and veterans of the B&O. Claude Crozet, as mentioned in the previous chapter, had taught at the Military Academy prior to Thayer's arrival in 1817. Lastly, Wilson C. Fairfax of Virginia had entered the Academy in 1816, but did not finish.[37] This mixed group of graduates, a former professor, and cadets who did not finish at the Academy, suggests that their education at West Point had provided them with suitable skills needed for engineering work. Leadership among the early professional engineers recognized the skill of those associated with West Point by having them comprise 40 percent of the constitutional committee. For the graduates and Crozet, the Military Academy imparted a common base of engineering knowledge. For Fairfax, who did not complete his studies, there was at least an attempt at formal education in science, mathematics, and basic engineering. In cooperation with non–West Point engineers, these men pushed to codify and distinguish the profession of civil engineering in the wake of the early railroad boom.

At the March 1839 meeting in Philadelphia, the members invoked a passage from Thomas Telford's 1820 inaugural address to the British Institution of Civil Engineers. Telford's remarks provided an important contrast between professionalism in Europe and in Great Britain:

> In foreign countries, similar establishments are instituted by government, and their members and proceedings are under its control; but here, a different course being adopted, it becomes incumbent on each individual member to feel that the very existence and prosperity of the Institution depend

in no small degree on his personal conduct and exertions, and merely mentioning the circumstance will, I am convinced, be sufficient to command the best efforts of the present and future members, always keeping in mind that talents and respectability are preferable to numbers, and that from too easy and promiscuous admission, unavoidable, and not infrequently incurable, inconveniences perplex most societies.[38]

Telford's passage highlights how the American society defined its own purpose and scope. Though Telford was referring to the British system of government and individual rights, the American engineers gathered in Philadelphia thought it applied to the United States as well, and read Telford's remarks for the record. Individual liberty may have been protected for all citizens under the Constitution, but not every citizen could be a "professional" in their chosen trade. For Latrobe and other members, the very nature of the relationship between the American government and the electorate demanded that true professionals police themselves. A professional had to be beyond political reproach, above party and faction. Not all craftsmen and artisans were qualified to be a part of the profession, and those who were had to meet the standards agreed upon by the collective whole. Through the construction of canals and railroads, the early engineers had developed a sense of stratification by "task, title, and income" and, more importantly, they sought to distance themselves "from such marginal types of engineers as mechanics, architects, toolmakers, inventors, contractors and scientists."[39]

In April 1839, the committee of seventeen drew up a proposed constitution, which required each member to make some written contribution annually or be fined ten dollars.[40] This provision and irreconcilable sectional differences among the engineers led to the proposal for four regional associations instead of one national organization. With that, the effort to create a national association of civil engineers languished for the next decade. Contributing to the lack of group cohesion was the actual geographic dispersion of ongoing engineering projects in the United States. Railroads, by design, mandated that engineers work over long distances to connect populated areas of the expanding nation. The career of an early engineer required mobility, and that mobility tended to impede regular attendance at professional meetings.

By 1850, the form, function, and caliber of an effective professional engineering society emerged in the Boston Society of Civil Engineers. Founded by James Laurie, an engineer who had emigrated from Scotland, the organization set strict requirements for membership. Given the concentration of engineers in Massachusetts, Rhode Island, and Connecticut, Laurie and his fellow engineers could require members to be mature, less transitory, and more prominent in their communities.[41] The railroad and bridge projects of New England in the 1840s and 1850s made it possible for engineers to maintain a more long-term presence in their communities, and more importantly, to participate in regularly scheduled meetings. Laurie's society, which occupied a permanent headquarters building with a meeting room and a library, became the model for the larger national society based at the Croton Aqueduct Department two years later.

GETTING CROTON WATER TO THE CITY, 1831–1852

The Croton Aqueduct (also known as the Old Croton Aqueduct) was the key project that attracted a cadre of engineers to New York City and kept them based there. First proposed in 1831, the project involved a campaign by the city's Common Council to bring the freshwater of the Croton River over a distance of forty miles from Putnam County to the center of Manhattan.[42] New York needed clean and plentiful water to stem waterborne diseases and to enable its firemen to put out fires. Croton water promised to mitigate these and other urban afflictions. Besides the legal, financial, and political challenges, the main physical hurdle of the project was carrying the Croton water over the rocky terrain of Westchester County and the Harlem River into the city. Colonel De Witt Clinton Jr., one of the sons of the famous New York governor, was the first engineer to plan for and initiate design of the waterworks.[43] Colonel Clinton (not a West Point graduate) proposed a three-year timeline that included surveys and construction of a gravity-fed channel to Manhattan at a cost of $11.5 million.[44] By the time the city leaders had secured the favor and funding to proceed with the project, the younger Clinton had died in Cuba in 1834 trying to recover from disease in a warmer climate. As the water commissioners would learn, the Croton Aqueduct was going to take much more time, personnel, and money than they had originally estimated.

Almost by default, in 1833 the Croton commissioners, and specifically Myndert Van Schaick, asked West Point professor David Douglass to be chief engineer. Van Schaick and the Water Commission were impressed by Douglass's reputation as a canal builder. Douglass was not a West Point graduate,

but as one of Thayer's former faculty members, he appeared more than qualified to devise a system to bring Croton water to the city. In his 1835 report to
the water commissioners, Douglass detailed the route, grading, structures, and
estimated costs to complete the aqueduct, a project he estimated would last
but four years.[45] For three years, Douglass led surveys and studies determining
the shape and route of the waterworks, but he failed to produce any substantial construction or physical progress.[46] Each year Douglass requested to hire
more engineers to assist him and, eventually, the Croton commissioners began
to question Douglass's engineering prowess, citing "a lack of energy in the
operations of their engineer department."[47] Douglass's time with the Croton
project proved disappointing for him and the Common Council. At the time
of his dismissal in 1836, only planning and surveys had occurred. To be fair to
Douglass, the aqueduct that was eventually constructed, especially the High
Bridge over the Harlem River, resembled his original design.[48] Additionally,
differences over the project's timetable between Douglass and the chairman of
the Water Commission contributed to his dismissal.[49] Douglass was better at
conceptualizing the building of the aqueduct than actually constructing it. As
he was more a theorist than practitioner, Douglass's tenure as chief engineer
employed by the city did not reflect the future success that some of his West
Point students would enjoy nearly two decades later.

Instead of relying on West Point–trained engineers to build the aqueduct,
Croton commissioners looked to engineers trained through the master and
apprentice system of the Erie Canal. After firing Douglass in 1836, Stephen
Allen and Van Schaick, two Democratic Croton commissioners, hired John
Jervis to be chief engineer of the aqueduct. Jervis's work on the Erie Canal
and its supporting canal network was proof enough to Croton officials that
he could succeed where the West Point professor had failed.[50] Jervis proved a
better engineer and administrator for the Croton project than Douglass had
been.[51] Even so, Douglass's four-year project turned into a ten-year ordeal for
Jervis's engineering team as they persevered through tense political pressure
and repeated economic perils. In the summer of 1842, Croton water reached
the north end of Manhattan via a series of engineering feats, including a 36-
inch pipe embedded in a coffer dam over the Harlem River.[52] On 14 October
1842, the dams, tunnels, aqueducts, bridges, reservoirs, and hundreds of miles of
pipes finally brought water from Putnam County to the fountain at City Hall
Park at a price tag of well over $12 million.[53] Amid the fanfare of that cloudy
autumn day, Croton water commissioner Samuel Stevens praised the "skill and

science" of David Douglass and the "performance of duty" of John Jervis.[54] Douglass's survey work and Jervis's construction were just the beginning of New York City's freshwater system. Maintenance and expansion of the Croton water system ensured that the Water Commission and, later, the Croton Aqueduct Department would employ a qualified cadre of engineers in New York through the end of the century.

Jervis remained as chief engineer of the Water Commission through 1848, when the New York State Legislature created the Croton Aqueduct Department. In accordance with the provisions of the newly created department, the mayor appointed Alfred W. Craven as chief engineer and commissioner of the department.[55] Born the son of a naval officer in 1810, Craven studied law at Yale and Columbia College, and passed the bar before being drawn to the life of a civil engineer in 1835.[56] Craven established himself as an engineer working on railroads in Ohio, South Carolina, New England, Pennsylvania, and New York.[57] When Craven took charge of the Croton offices at Rotunda Park in 1849, he occupied a municipal building that became the venue for his fellow engineers to meet and discuss their profession. No longer working as a transient railroad engineer, Craven made the Rotunda Park offices a professional home for civil engineers in the city and surrounding area.

Among his colleagues was George S. Greene. Craven and Greene had begun their professional relationship in 1837 in South Carolina while both men worked on the Louisville, Cincinnati, and Charleston Railroad.[58] They continued to consult each other in their various engineering enterprises. In 1841, for example, while surveying the Cumberland River basin for coalfields in western Maryland, Greene shared the details of his survey and land speculation with Craven.[59] From Maryland, Greene moved to New England to work for the Boston and Providence Railroad in 1849, the Kennebec and Portland Railroad in 1851, and the Providence and Bristol Railroad in 1853.[60] It was during this period that Greene and Craven joined their fellow engineers seeking to form a national organization of professional civil engineers.

Greene was the first West Point graduate to play a major part in the construction of the Croton Aqueduct. As a cadet from 1819 to 1823, Greene went through the Military Academy just as Sylvanus Thayer was establishing the Thayer method and formalizing the science and engineering curriculum. Prior to arriving at the military school, Greene had attended Brown University in 1817; however, the economic aftershocks of the Embargo Act of 1807 and the War of 1812 finally forced his father's Rhode Island shipping company to fail by

1817. Without money, Greene could not continue at Brown.[61] He went to work for a dry goods merchant in New York City and then secured an Academy appointment in 1819, determined to make the most of the free education and opportunity to be an officer. Douglass taught him mathematics, and Crozet was his engineering professor. Of the seventy-nine cadets that entered the class of 1823, thirty-five graduated, with Greene finishing second in the class. Typical for the top Academy graduates during the Thayer years, Lieutenant Greene stayed on to teach mathematics for four years.[62] After West Point, Greene served eight years at Fort Sullivan in Maine. While there, he lost his first wife and three children to sickness over a seven-month period. This personal tragedy led Greene to look beyond the army for a new start. He resigned his commission at the end of 1835 and started his engineering career working on a railroad in Massachusetts. When he met Craven in South Carolina in 1837, Greene's engineering record consisted of eight years at West Point as a cadet and instructor, and two years of civilian experience.[63] Craven made sure that Greene was invited to the first meeting of the ASCE in 1852, and in 1856, Craven sought out Greene to build a new receiving reservoir in Central Park for the Croton water system.[64]

Another West Point graduate who worked on the Croton Aqueduct was William H. Sidell, class of 1833. He was also one of the founding ASCE members in 1852. Sidell pursued a civil engineering career by resigning his commission almost immediately after graduating sixth in his class, two places below the cutoff to be commissioned into the Army Corps of Engineers.[65] Looking to capitalize on his civil engineering education, Sidell spent four years working in New York City. First as a surveyor, and then as an assistant engineer of the Croton Aqueduct, Sidell worked for Douglass in the autumn of 1834, surveying the Croton River valley.[66] Later, Sidell served as a railroad engineer and an assistant engineer on a dry dock project in New York Harbor.[67] From New York, he traveled west in 1840 to play a role in the hydrostatic survey of the Mississippi River delta, and spent the next twenty years as a railroad survey engineer in the northeastern United States, the United States west of the Mississippi, and Mexico.[68] When Sidell became the seventh member of the ASCE on 1 December 1852, he was an engineer for the Isthmus of Tehuantepec Railroad in Mexico.[69]

For both Greene and Sidell, New York City was a steady source of income and status for their engineering careers. The Croton Aqueduct project, the growing fraternity of engineer peers, and the ongoing transition from canals to railroads all made New York City the hub of engineering in the antebellum era. The world took notice of the Croton achievement, wondering how in "a country which was every day representative in a bankrupt, hopeless condition [that] so great and expensive a work should be brought to a conclusion." Observers wondered how a developing nation could build the High Bridge over the Harlem River.[70] In London, one reporter called it "one of the most stupendous works of modern times."[71] It was no coincidence that American engineering located its professional institution in New York at the same time the "Emporium City" was becoming the financial, social, and political center of the United States, especially between 1844 and the Civil War.[72] Wall Street was the key to American investment in westward expansion, and engineering expertise gravitated to the financial capital.[73] While engineers under Douglass and Jervis were building the aqueduct, New York City, propelled by the stimulus of trade from the Erie Canal, outpaced its rivals in commercial growth. By the end of the 1840s, the city was the center of a "network of exchange and interdependence which drew Americans out of local isolation and into a modernizing society to the general economic benefit of all and the special benefit of New York."[74] During House debate over an 1851 mint bill, one Indiana congressman enviously lamented that "[the] city of New York controls at present time, with its immense monetary power, the commercial destinies of the Union."[75] For mid-Victorian engineers such as Sidell and Greene, New York promised prestige as well as financial security at a time when boom, bust, and panic were economic facts of life.

Given the salaries the city government paid its appointed engineers, this time of modernization also had special benefits for men with the title of "engineer-in-chief." Douglass and Jervis each received an annual salary of $5,000 from the state while heading Croton construction in the 1830s.[76] These salaries were equivalent to what the engineer of a large project, such as a railroad, might earn.[77] As chief engineer of the Croton Aqueduct Department, Alfred Craven received an annual salary of $2,000, the same as that of the board president and commissioner.[78] Assistant engineers and survey engineers could earn between $800 and $1,000 per year. During the canal and railroad boom of the 1830s, American engineers' salaries generally were the highest on a project's payroll up until completion of construction, as the salaries of Douglass and Jervis were for the aqueduct. At the time of project completion, the chief engineer's employment ceased, and employment of the maintenance or "resident engineer" began with an annual salary of $2,000, similar to what Craven received from the Croton Aqueduct Department. In antebellum New York, engineers had opportunities for both types of employment in one location. With the varying income potential that could come from the titles of chief engineer, ordinary engineer, and engineer of second rank, distinguishing a civil engineer from a builder, artisan, or mechanic became paramount, especially in a growing metropolis like antebellum New York.

FOUNDING OF THE AMERICAN SOCIETY
OF CIVIL ENGINEERS, 1852

According to the national census of 1850, there were 512 engineers in the United States, with the largest concentrations in Massachusetts, New York, Ohio, Pennsylvania, Connecticut, and Wisconsin. Of the 512, there were 58 (11 percent) West Point graduates working as "civilian engineers," while 634 of 997 living Academy graduates were still in the army.[79] Whether a West Point–trained engineer, a graduate of a civilian school, or a product of a master and apprentice system, the mid-century engineer was independent, respected, and well paid. Engineers saw themselves as protecting the public interest and possessing integrity beyond reproach.[80] The prestige and wealth attained by engineers, especially during the railroad boom, heightened this sense of importance and professional identity. Encouraged by the success of the Boston Society of Civil Engineers, James Laurie lobbied his New York colleagues to try again for a national organization.

On 23 October 1852, Craven, along with five other New York civil engineers, sent out a formal invitation for the first meeting of the American Society of Civil Engineers and Architects. Two weeks later, on a cold Friday evening, twelve men convened the society's first meeting in Craven's office near City Hall. In addition to the five signers of the invitation, present were Thomas A. Emmet, J. W. Ayres, Edward Gardiner, Robert Gorsuch, George S. Greene, Simeon S. Post, and W. H. Talcott.[81] Greene and Sidell were the only West Point graduates there, but Julius Walker Adams had attended West Point for a year in 1830–1831 with the class of 1834.[82]

That evening, the members voted for and approved the society's constitution. The twelve engineers agreed that the society's objectives were the "professional improvement of its members, the encouragement of social intercourse among men of practical science, the advancement of engineering in its several branches, and of architecture, and in the establishment of a central point of reference and union for its members." Additionally, the new ASCE constitution noted that New York was the center of "commercial importance." The society sought to be a forum where engineers from multiple specialties could interact intellectually and socially.[83] Architects were always included in the society, but their name was dropped from the its title in 1868.[84] In its constitution, the society recommended several means to accomplish its objectives, including "periodical meetings for the reading of professional papers, and the discussion of scientific subjects, the foundation of a library, the collection of maps, drawings and models, [and] the publication of such parts of the proceedings as may [be] deemed expedient."[85] Among officers elected at the first meeting was Laurie as president. Among the five directors selected was one West Point graduate, Sidell.

Similar to its predecessors, the 1852 version of the American Society of Civil Engineers was slow to take off. Over the course of the next two years, the organization held only fourteen meetings, with an average attendance of six members.[86] Membership did expand to forty-eight by the end of 1853, and increased slightly to fifty-four members in 1854. Eleven of the original members had attended West Point. The society collected dues from members in New York and beyond, with the New York members paying higher fees.[87] By charging non–New York residents less for membership, the Board of Directors hoped to attract more engineers from around the country to join the society. In their first annual report of 1853, they made six engineers honorary members, including four West Pointers: John James Albert, Alexander Bache, Dennis Hart Mahan, and Joseph Totten.[88] The society's recognition of the honorary members from West Point reflected the significance of the Military Academy to the "eminence" of the engineering profession in 1853.[89]

Fulfilling the intent of the society's constitution turned out to be more difficult in practice than in theory. Member attendance diminished in spite of the professional presentations given at several meetings. The offices of the Croton Aqueduct Department soon became inadequate as a permanent home for the society. Alfred Craven's Rotunda Park office was part of the city's administrative facilities, and there was no way to set aside space for a permanent library and meeting room for members to exchange ideas.[90] The ASCE of the 1850s was still a regionally focused organization, accommodating the wants and will of the New York engineers. Nonresident members could have been better incorporated into the society through publications and circulars, but the early

ASCE did not publish any proceedings that could have been beneficial for members outside of New York to read. In March 1855, the society recognized these shortcomings and transitioned to an inactive status for the next twelve years. James O. Morse, the secretary, remained the lone officer in place from 1855 to 1867. The ASCE's inactivity was largely due to Laurie's departure from New York for employment on new railway projects in Nova Scotia and New England. As ASCE president, Laurie had ensured that the members met on a regular basis. Without Laurie, the organization lacked leadership and focus. External events also overwhelmed the nascent professional organization's efforts. The outbreak of the Civil War further diverted attention and also contributed to the inactive decade of the ASCE.[91]

Still, the founding of the ASCE meant another expression of citizenship in the antebellum United States. ASCE members' values reflected an urban ideal of republicanism, where sovereignty resided with the people. Just as infrastructure and municipal projects sought to create order in a disordered urban landscape, these engineers organized to promote self-discipline, virtue, and morality in the democratic civilization they were building.[92] Lofty though their goals may have been, the reality of the unresolved questions of the American Revolution, specifically slavery, proved to be beyond the scope of what the engineers could grapple with in the 1850s.

GREENE AT THE CROTON AQUEDUCT DEPARTMENT, 1856–1862

Unfortunately for George S. Greene, the ASCE went inactive at the same time Alfred Craven convinced the Croton Aqueduct Department to bring Greene to New York to expand the Croton water system and build the new reservoir in Central Park. Greene, living in Rhode Island, had been one of the non–New York residents supporting the society. Prior to 1856, he attended ASCE meetings when his travels to New York permitted, but his attention remained focused on building the Providence Railroad.[93] Greene's connection with Craven and his stature as an engineer made Greene a favorite of the Croton men. Between 1856 and 1862, Greene designed and constructed the extension of the city's water supply. The highlights of Greene's work consisted of the "large distributing reservoir in Central Park, 88 ft. deep, and covering 96 acres; the construction of a wrought-iron pipe, 90.5 ins. in diameter and 1,400 ft. long on

High Bridge, across the Harlem River; and the laying of a cast-iron pipe, 60 ins. in diameter and 4,116 ft long across Manhattan Valley." According to the ASCE memorial to Greene, his "reservoir and pipe construction"were the first of their kind, minimizing leakage and water loss by laying "trenches in the solid rock."[94] Greene's ideas were innovative in the engineering field of 1858 and were copied in water projects under construction in other American cities.[95]

Greene's major contribution to the city's landscape was leading the construction of the New Reservoir in Central Park. Completed in August 1862, the New Reservoir became one of the main features in the Central Park design, known as the Greensward Plan of Frederick Law Olmsted and Calvert Vaux.[96] However, the New Reservoir construction project was separate from construction of the park, which created some controversy.

When Republicans in the New York state legislature reorganized the city government and undercut the authority of the Common Council in the charter of 1857, they cleared the way for Olmsted and Vaux to lead the building of Central Park. To many in Albany, Fernando Wood and the Tammany Democrats had made Central Park a hub of corruption and patronage. The Croton Aqueduct Department remained unchanged by state Republicans because the engineers there appeared untainted by Wood and the Tammany leadership. Moreover, the Croton engineers, led by Craven, made their case for state control of the planning, construction, and management of the water supply and sewers.[97] Other than agreeing with the Croton Aqueduct Department as to the general shape of the reservoir, the Central Park commissioners and Olmsted had no control over construction of the New Reservoir.[98] Greene reported to Craven, not Olmsted, and thus had independence in the direction and scope of the project. At Greene's recommendation, the Croton Aqueduct Department selected the firm of Fairchild, Coleman, Walker & Brown as builders for the reservoir. The Common Council approved the action and permitted Greene and the firm's men to break ground in April 1858. Greene supervised a team of civilian engineers, 1,200 laborers, and more than one hundred horses.[99] For nearly four years, the park commissioners and Olmsted repeatedly lamented that they had no control over the construction of the new lake, leaving the structural details, grading, and building responsibilities to the Croton Aqueduct Department and Greene.[100] With Craven as head of the Croton Aqueduct Department, Greene did not have to concern himself with the political maneuvering of Tammany Democrats, Fernando Wood, and the Republican state legislature. However, the national crisis over slavery and states' rights was too great for Greene to ignore. Just as the Civil War had interrupted the momentum of the American Society of Civil Engineers, the war also suspended Greene's career as an engineer with the Croton Aqueduct Department and the city.

During the nation's descent into war, Greene had remained mostly removed from any formal allegiance or political activity other than voting. As his son recalled in 1902, Greene's "sympathies were with the Whig party" in the 1850s, a likely position for Greene given that Whigs supported funding for internal improvements.[101] Greene was not an abolitionist, as were some of his family members, but he did have a strong sense of duty to defeat the rebellion and preserve the Union. At the age of sixty, Greene, like many of his fellow West Point graduates, sought to reenter the army and defend the United States against the Confederate rebellion. Although he had been out of active service for over twenty-five years, he secured a commission as a colonel with the 66th New York Volunteers. Greene left the Croton Aqueduct Department in January 1862 and took command of a regiment of volunteers in Maryland later that month.[102]

The Civil War experience of the West Point graduates added another dimension to their expertise and professionalism in New York. Once war broke out, the men demonstrated their patriotism and loyalty to the Republic. Greene is but one illustrative example; his Civil War exploits are well chronicled in military histories.[103] Greene led his men at the Battles of Cedar Mountain, Antietam, and Fredericksburg in 1862. In July 1863, he defended the Union's right flank on Culp's Hill during the Battle of Gettysburg. At Culp's Hill, Greene made his legendary stand that saved Gen. George Meade's headquarters, and directly led to the Confederate defeat at Gettysburg.[104] From Gettysburg, Greene fought at Chattanooga and in North Carolina before taking part in the Union Army's grand victory review parade in Washington on 25 May 1865. What is remarkable about Greene's Civil War service was not only his advanced age, but what he had to endure. At Lookout Mountain in 1864, a bullet passed through his cheek and lower jaw, debilitating him for several months. Greene also had two horses shot out from underneath him during the course of fighting.[105] By war's end, he had advanced to the rank of major-general by brevet. Already a proven professional engineer, he emerged from the war a decorated survivor and a celebrated hero of the Union. As would be the experience of so many West Point veterans, his hero status and accolades led to greater prestige and access in the elite class of postwar New York.

The Civil War also heightened attention to improving sanitary conditions in American cities. Reformers in the U.S. Sanitary Commission urged state and local governments to establish organizations to could continue the public health efforts started in army camps and the occupied South. Among the first to respond, New York State passed the New York Metropolitan Health Law, which created the Metropolitan Board of Health for New York and Brooklyn in 1865.[106] The Citizens' Association spearheaded the call for health reform in New York, realizing that filth and unsanitary water sources created the "evil circumstances" prevalent in the city. More importantly, the 1866 report of the Citizens' Association argued that the welfare and health of all classes of people, as well as "Sanitary Science," outweighed "the physical, social, political, or commercial interests of the city."[107] After the war, General Greene benefited from this newfound emphasis on the value of public welfare and health.

GREENE IN POSTBELLUM GOTHAM

The U.S. Army released General Greene from service on 30 April 1866. While the Civil War ravaged the South and a generation of Americans, New York City and the Croton Aqueduct Department continued to improve and expand its water system. Based upon the previous survey work of Douglass and others, the department identified which parts of the Croton River could be dammed to create more reservoirs in the future. On St. Patrick's Day, 1866, the department decided to build the first expansion reservoir "at Boyd's Corner on the West Branch of the Croton Valley."[108] On May 1, the department brought back Greene for this project. That Greene was now a hero as well as an accomplished engineer made him an easy choice for the department. At Boyd's Corner, the commissioners sought to build a dam without precedent in the United States. Greene designed a dam 78 feet high and 670 feet long, with the capacity to retain over 2,700 million gallons of water.[109] Drawing upon his experience with construction material and his education, Greene mastered the problems presented by the prospect of building the new dam in Putnam County. By the end of August 1866, Greene had drawn up the plans for the dam at Boyd's Corner, enabling the Croton Aqueduct Department to award the contract and start work.[110]

The dam at Boyd's Corner took nearly seven years to complete. While it was under construction, several important changes occurred in the Croton Aqueduct Department and in Greene's employment with the city, all stemming from William "Boss" Tweed's consolidation of power at Tammany Hall. First, in May 1868, after twenty years of service, for health reasons Alfred Craven retired as the department's chief engineer, and Greene took his place. Greene's promotion was temporary.[111] During 1869 and 1870, the Tammany Democrats took back control of the municipal government, with the state legislature passing the 1870 city charter. With A. Oakey Hall as New York's (and Tammany's) mayor, Tweed merged the Croton Aqueduct Department and the Streets Department to form the Department of Public Works in 1870.[112] Under Craven, the scope of the Croton Aqueduct Department's responsibilities had expanded from delivery of water to the city to include the laying of new sewer and gas lines. Here, the engineers' desire for centralized control in the construction of sewer and water infrastructure merged with Tammany's desire to control all public works and services in the city. Tweed and his fellow Tammany cronies saw the patronage potential of the Croton Aqueduct Department and ensured that all additional "responsibilities" fell under the Department of Public Works. Led by Tweed, the new department became a source for Tammany patronage and grand city projects.[113] Ever the professional engineer, and perhaps conscious of the potential tarnish to his character that could come from being associated with Tammany's machinations, Greene resumed leading engineering projects as directed by Tweed and then the city after Tweed's demise in the autumn of 1871.[114]

From 1 May 1870 until 11 January 1871, Greene served as an assistant engineer in the Department of Public Works. From 1868 to 1871, he also served as a consulting engineer on the Morrisania Survey Commission, which provided a "system of exact topographical surveys and monumenting of base lines and of street lines" in New York City. After a short stint designing sewers in Washington, DC, Greene came back to New York as a consulting engineer to the Department of Public Works from October 1872 to September 1873. There he consulted on the construction of new bridges and tunnels across the Harlem River.[115]

Throughout his time with the Croton Aqueduct Department and with the Department of Public Works, Greene was in every respect the skilled engineer. He maintained carefully handwritten notebooks, complete with schematic cross sections of the aqueduct's pipes and dimensions. For nearly two decades he recorded calculations, mathematical equations, flow rates, and aqueduct drawings in a 154-page personal notebook. Greene recognized that he needed to work with other city commissions, especially the Central Park Commission, as he expanded the Croton system. On the last page of his notebook he tracked thirteen "Central Park reports" from 1857 to 1869, ensuring that he had a current list of references for his projects. The other telling revelation from Greene's notes was that he recognized the need to understand what his predecessors had done in making the aqueduct system. In several pages, he summarized Jervis's works and the reports of the Croton water commissioners prior to the formation of the Croton Aqueduct Department.[116] Greene's notes and annotations were handwritten much more carefully than his correspondence. Reading his engineer notebook, one quickly understands the pride and sense of professionalism Greene had as a civil engineer. Starting the notebook before the Civil War and having it with him when he returned to New York in 1866, he was quickly able to resume his engineering work.

Missing from Greene's records are any references to political developments or hint of practicing political patronage. He did not appear to be drawn to the intrigues of either political party in the 1860s. When Tammany Hall fell in 1871, Greene was not connected in any way with Tweed's circle of corruption. Of the West Point men who came to New York City, Greene was the ideal. Serving his country in war and the engineering profession in peace, Greene was a commendable graduate.[117] Maintaining his reputation and profession in the age of Tammany also made Greene an exemplary member of the American Society of Civil Engineers. His status as a hero of Gettysburg helped him to remain free of innuendo and scandalous accusations. Upon Greene's death in January 1899, both West Point and the ASCE published memorials eulogizing his life as an engineer and Civil War hero. The *New York Times* noted that through his study of mathematics Greene had "attained a degree of proficiency that placed him in the ranks of experts," and through his "brave and efficient service" he became "major general" in the army.[118] With a lifetime of service to the profession and the nation as an army officer, as well as his long list of engineering accomplishments in the city, Greene was the model of professionalism for the ASCE. His wartime exploits poised him to resume his place as a trusted professional engineer for the city, and start the next chapter in the ASCE.

THE ASCE REVIVAL

Just as Greene had his notebook to help him pick up where he left off before the Civil War, the ASCE had James Laurie. In 1866, Laurie went to New York, and with James Morse, the lone remaining officer of the society, resuscitated the American Society of Civil Engineers. Laurie called for a meeting at the office of C. W. Copeland, 171 Broadway, on 2 October 1867. Laurie, Morse, James K. Ford, William J. McAlpine, Israel Smith, and McRee Swift agreed to reorganize the ASCE.[119] The outcome of that meeting was a committee report that addressed how the organization would renew activities and revitalize itself. Without dwelling on why the ASCE had been inactive for a dozen years, the men looked forward to making it permanent as they rededicated themselves "to science and to art." Recognizing that the majority of members were going to be "non-residents of New York," they acknowledged that the society needed to be "a fixed institution."[120] To that end, the New York men decided to procure a permanent home for the organization, on the corner of William and Cedar Streets in the vicinity of Broadway and Wall Street (63 William Street).[121] With renewed energy and a building of its own, the ASCE grew quickly from the surviving thirteen members of 1855 to over 179 members by the end of 1870.[122]

The other impetus for reviving the ASCE in 1866 was the postwar migration to New York of men claiming to be engineers, or at least those possessing some expertise in engineering gained through their wartime experiences. In 1867, J. P. Kirkwood was elected president of the ASCE and Julius Adams became vice president. Under their leadership, the society limited membership to "accomplished and competent men." Qualifications for joining the society were at least five years of supervisory experience as an engineer, including military engineering service, or the completion of an engineering program at a college such as Rensselaer Polytechnic Institute or West Point.[123] The intent of these membership guidelines was to distinguish experienced professional engineers from laboring-class workers and mechanics who lacked the education and experience to lead a project responsibly. Since engineers were part of the capitalist bourgeois elite in the postwar city, the ASCE needed to manage growth and exclusivity simultaneously in order to preserve its position and status.[124] In an 1887 study, the ASCE noted that its greatest growth in membership corresponded to "great periods of prosperity" and that it showed the least growth in "years of great depression." This was certainly the case after the Civil War. To accommodate this period of "prosperity," in 1877 the society stratified its membership into "members, juniors, associates, and others" as a way to increase future expertise and expand the power of the organization.[125] By creating varying levels of membership for varying levels of engineering expertise, the ASCE nearly doubled its total membership from 552 members in 1876 to 1,019 members in 1886. As a result, the ASCE remained the leading organization of professional engineers through the end of the nineteenth century.

The other challenge for engineering professionals in postwar New York was breaking the bonds of political favoritism. An example of such favoritism occurred in the summer of 1860, when Mayor Fernando Wood attempted to fire Craven as chief engineer of the Croton Aqueduct.[126] In that episode, Wood accused Craven of incompetence and corruption in order to remove him from the project and to reassert municipal control over the state-legislated Croton Aqueduct Department. Wood wanted one of his patrons, Cummings & Co., to receive lucrative contracts for the new reservoir even though that company's bid was not the lowest or most suitable submitted to the department.[127] Craven responded with an eighty-page explanation to the Board of Aldermen, explaining that he was "obliged to go into details which would be unnecessary if [he] were addressing a body of engineers."[128] In defending his position, Craven relied on his professionalism and technical competence to refute the charges of the Democratic mayor. The chief engineer also referred to the West Point credentials of fellow engineer "Captain George S. Greene" as he explained why the new reservoir construction required more rock removal than they had estimated.[129] Craven successfully fended off Wood's charges and continued his tenure as chief engineer until his retirement in 1868.[130] Upon retirement, Craven focused his energies on the ASCE, serving as its president from 1869 to 1871 and acting as an engineer consultant on several public works projects in New York and Georgia.

Similar to what the ASCE would do for George S. Greene in 1899, the society canonized Craven as the epitome of the civil engineer in 1879. Craven, however, rated higher in the pantheon than Greene. Between 1839 and 1879, Craven was present at every major meeting in the engineers' efforts to form a viable professional organization. From Augusta to Baltimore to New York, Craven improved his credentials working on railroads, port facilities, and waterworks. For Craven, being an engineer entailed moral obligations beyond any state or federal law. In his opinion, the engineer served the public to the fullest of his duty to protect "their rights and in the preservation of the integrity of the Department on which they rely."[131] When Craven died in 1879, the society saw him as the ideal professional engineer who possessed the expertise, integrity, and character desired by the profession.

For the ASCE, George S. Greene was a close second to Craven as the benchmark of professionalism in engineering. Greene spent the nineteenth century "cautious and conservative" in his political engagement, but he was always forward-looking in his profession.[132] He avoided being drawn into Tammany's web under Wood in 1860 and under Boss Tweed in 1870. In the 1860 incident, Craven's testimony before the Board of Aldermen had protected Greene from any suspicion of corruption. Boss Tweed simply abolished Greene's position in 1870, a clear sign that Greene did not benefit from Tweed's largesse.[133] Like Craven, Greene also continued to work as an engineer consultant in the decade following his termination with the city. As the venerable professional, Greene consulted on railroad and waterworks projects in New York, Washington, Detroit, and Providence. He also served a single term as president of the ASCE in 1876.[134] Under Greene's leadership, the society's membership grew to over four hundred members, with 70 percent of them living outside of New York City.[135]

Unlike Craven, Greene worked well into his eighties. In 1886, the Aqueduct Commission appointed him as "one of a Board of Examining Engineers to investigate charges which had been made affecting the management and the condition of the work which had been done." As old as he was, Greene "insisted on walking through the entire length of the tunnels, examining closely everything as he went, a task to which his [younger] associates . . . found themselves unequal." This was Greene's last "official" act as an engineer.[136] Greene lived to the age of ninety-eight, a man celebrated for both his military valor at Gettysburg and a highly commendable life as a professional engineer. His longevity alone added to his fame among the nearly 2,200 members of the ASCE of 1897.[137] The ASCE histories and proceedings published at the end of his life and after his death celebrate Greene among their vaunted group of great engineers. The accuracy of these memorials does not matter as much as the ideals and symbolism of Greene's narrative to the profession. In an era when professions and professional associations flourished, the ASCE had Greene as an example to show the value and impact of engineering to its peer professions.

Craven may have been a model of integrity for engineers, but Greene was the ideal for dedication to and resilience of the profession. Greene was progressive in adapting new technologies and techniques to his projects, particularly with the construction of dams and water system infrastructure. In his later years, he advocated for the implementation of an underground rapid railroad system to alleviate traffic congestion in Manhattan.[138] Always looking to improve life for New Yorkers, Greene consulted engineers on building and expanding sewers in the Bronx, Brooklyn, and Yonkers. Greene remained consistently devoted to the principles of the ASCE and the ideal of the professional.

Like many other graduates in their last years, Greene returned regularly to West Point. During the 1880s and 1890s, he was the oldest living graduate and thus spoke at the annual reunion of Academy graduates held each June. Reflecting upon his alma mater in 1888, Greene recalled: "The officers and graduates, by their steadfast devotion to duty, have placed this Academy in a career of usefulness and of honor which has brought to it the high appreciation and hearty support of the Government and the people; the attacks and opposition of the past generation have been succeeded by esteem and good will. We shall look with confident hope to the succeeding officers and graduates to maintain this honorable record."[139] Most would rightly assume that Greene was thinking of military service in his remarks. But given his experiences as an engineer before and after the Civil War, "usefulness and honor" applied to the civilian pursuits of each graduate as well. The profession of civil engineer, as espoused by Greene and his fellow ASCE members, aspired to be the most useful and honorable profession in the United States.

CONCLUSION

Greene, Sidell, and, to a lesser degree, Douglass used both the technical and moral aspects of Thayer's West Point system in their civilian careers. Encouraged by the Survey Act of 1824 and the westward expansion of Jacksonian America, these graduates validated the discipline and curriculum of the cadet experience by building canals, railroads, and waterworks in the United States. Graduates of the Military Academy possessed the self-discipline and technical capacity to work with master-and-apprentice civilian engineers. At the same time, civilian engineers championed their West Point colleagues' education and expertise as proof that engineering was a legitimate, important, and demanding profession. For the engineers in New York City, adhering to these principles of duty, self-discipline, and service enabled the ASCE to grow and thrive, especially after the Civil War. However, as Craven's and Greene's experiences with Tammany officials showed, there were limits to the influence of a true professional. Both men steered clear of Tammany, Tweed, and the potential to tap into the spoils of the Public Works Commission. While largely successful in their engineering exploits, the earlier generation of West Point engineers did not expand their status into the business bourgeois elite that emerged in the 1880s.[140] The next generation of West Point men to come to the city would not be as apolitical or as hesitant to join the business elites.

Even though there were limits to the power of professional engineers in antebellum New York, these men enabled the city to accommodate the commercial boom and the population explosion before and after the Civil War. Croton water, city sewers, and bridges and railroads over the Harlem River made the metropolis not only livable, but also desirable. New Yorkers acknowledged this in celebrating the opening of the aqueduct and the expansion of the new reservoir, and in their deference to the Croton engineers during the Wood scandals of 1860. City leaders' expectations and the ability of their engineers to meet or exceed those expectations set professional engineers at the top of the middle class in terms of both economic rewards and social status. West Point graduates who fought in the Mexican War saw the success of older alumni in the city, and would take their chances as well.

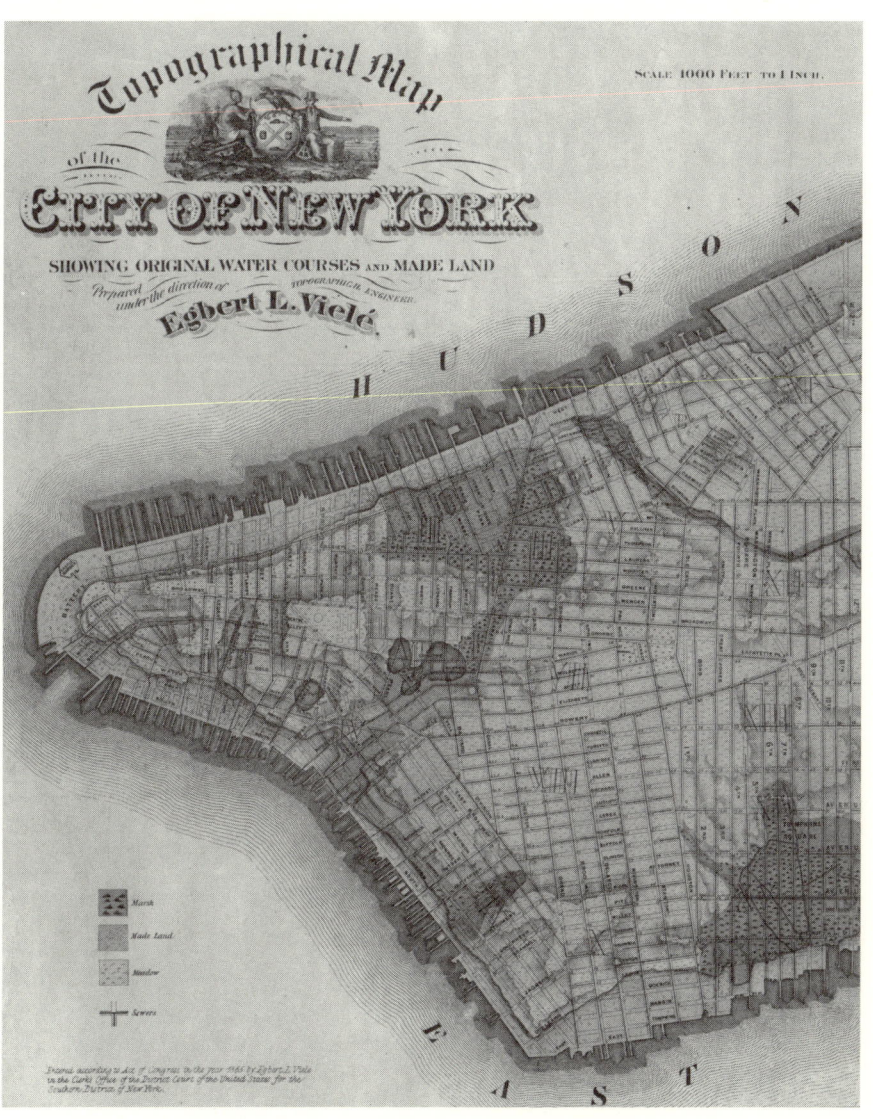

Detail of *Sanitary & Topographical Map of the City
and Island of New York,* known as the Viele Water Map, 1865

(Library of Congress, Geography and Map Division)

Brig. Gen. Egbert L. Viele, 1861

(Library of Congress, Prints and Photographs Division)

Maj. Gen. John Newton, 1862

(Library of Congress, Prints and Photographs Division)

Fitz John Porter, 1862

(Library of Congress, Prints and Photographs Division)

Henry Warner Slocum, 1869

(Library of Congress, Prints and Photographs Division)

Maj. Gen. George B. McClellan and his wife, Ellen Mary Marcy, 1862

(Library of Congress, Prints and Photographs Division)

Corps of Cadets marching in the Dewey parade on Riverside Drive, 1899

(Library of Congress, Prints and Photographs Division)

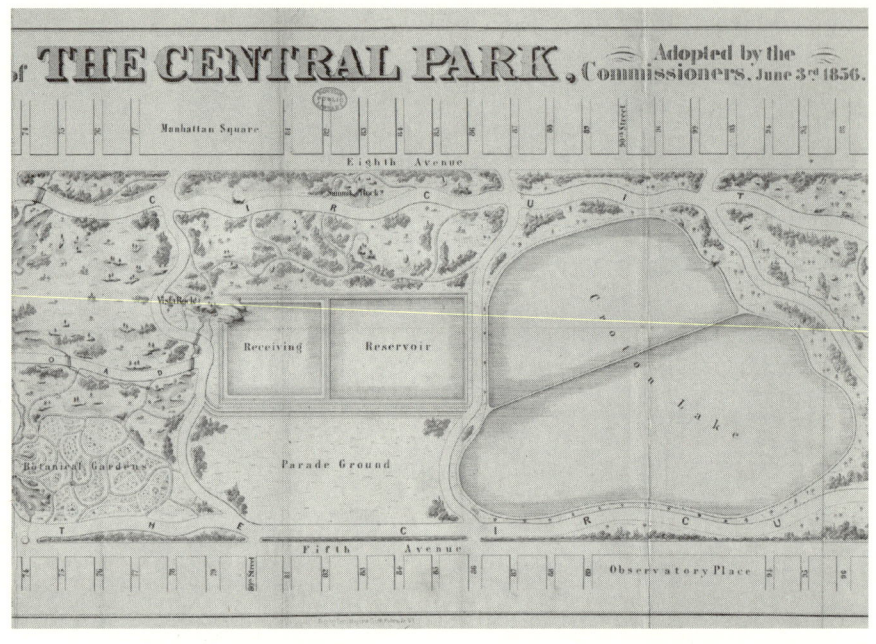

Detail of *Plan for the Improvement of the Central Park,*
by Egbert L. Viele (New York: Ferd. Mayer & Co., 1856)

5

EGBERT L. VIELE'S
NEW YORK

Let the work of improvement be begun at once, and those who conceived this measure will see it completed. The hot days of midsummer will soon return, with the pestilence in their train, and the overworked inhabitants will seek in vain a spot where they can breathe the pure air of heaven.

—EGBERT L. VIELE, *First Annual Report on the Improvement of the Central Park,* 4 January 1857

Neither man realized it at the time, but 11 September 1857 was the start of a personal rivalry between two prominent personalities that lasted through the end of the century. On that Friday in Central Park, Egbert Viele, the park's engineer-in-chief, met Frederick Law Olmsted, who introduced himself as its new superintendent. Olmsted reported to the chief engineer by pushing past a line of unemployed men standing outside the chief's rustic park office. All in line carried "letters" of endorsement, signifying to the chief engineer that they had sought their places with the imprimatur of Tammany and Mayor Fernando Wood.[1] But unlike the others, Olmsted's letter was signed not by Wood, but by another powerful Democrat on the park commission, Andrew Haswell Green.[2] After Viele ignored Olmsted for about half an hour, he dismissively told Olmsted that he preferred a "practical man" to be his superintendent and then sent the well-dressed Olmsted on a muddy inspection of workers clearing the park. Olmsted had not anticipated being sent directly to work upon that initial meeting. Recalling that first day in Central Park, Olmsted noted that there were about five hundred workers organized in gangs of fifteen. Every man was a Democrat appointed through the spoils of

Wood's patronage, and none seemed to pay much respect or mind to their new superintendent.[3]

Viele's less than hospitable reception of his new superintendent was understandable. For nearly two years, Viele had been supervising teams of immigrant laborers in clearing rocks and burning brush from the 800-acre swath designated to be Central Park. Mayor Wood had appointed Viele, West Point graduate and veteran of the Mexican War, as the engineer-in-chief to construct Central Park in accordance with the plan that Viele had drafted. But there were critics of Viele's plan and, earlier in the summer, they and Olmsted had schemed with several Republican friends to secure Olmsted a position supervising the construction of Central Park. While the Republicans wanted to stem the political power of Tammany, Olmsted and his partner, Calvert Vaux, wanted to ensure that Viele's plan, which they thought was substandard, could be stopped before it was too late. Through his acquaintance with Charles Elliott, a Republican appointed by the state to be one of the park's commissioners, Olmsted became superintendent and began a process intended to extract Viele and his Tammany patrons from the most substantial project led by the city's government up to that date.[4] Ultimately, Olmsted and Vaux succeeded in supplanting Viele and his plan for the park, but Viele's contributions would prove nonetheless to be critical to the creation of Central Park.

The political circumstances of Olmsted's and Viele's first meeting in 1857 emanated from the range of democratic manifestations that emerged in New York just before the Civil War and lasted through the end of the century. Even though Viele and Olmsted were rivals, both men pursued their endeavors under the mantle of reform. Viele, a Democrat, and Olmsted, a Republican, both claimed that they were addressing the ills of the modern city, but they did not agree upon the ways and means to achieve a better, safer, and more just New York. Their rivalry illuminates the nuances of the radical and conservative political factions competing for power there. Specifically, Olmsted was a leading "radical" figure, seeking to overthrow the status quo of Tammany power. Although he was not a Tammany Democrat, Viele more often than not represented the Democratic vision of shaping and expanding New York by adhering to more "conservative" tendencies.[5]

Disparate opinions over slavery and abolition provided another important contrast between Olmsted's emergent party of Republicans and Viele's New York Democrats. Racism infused much of the debate on both sides. Abolitionists advocated for the end of slavery, but not necessarily equality for African

Americans. Free labor ideology, an antecedent component of the Republican Party, opposed slavery not necessarily for moral reasons, but rather, to encourage the promise of new opportunities for free labor in the new territories and states. The influx of Irish and German immigrants into the New York demographic landscape fueled proslavery sentiments in 1850s New York. From 1830 to 1850, the percentage of black servants in Manhattan decreased from one-half to less than 10 percent. Irish maids worked in the homes of the elite while Irish labor controlled the docks and drayage trades. Dominated by nativists, and then white European immigrant working groups, New York Democrats sought to secure power through Tammany patronage and labor support in the wards of Manhattan. Particularly under Fernando Wood, the Tammany Democrats championed city labor and did not see African Americans, free or slave, as equals. In emphasizing labor and industry, and protecting the poor, Wood played class politics to solidify his power as mayor.[6] Against this backdrop of abolitionist virtue, racist opinions, and labor practicalities, Viele had ended up in the Democratic camp and Olmsted in the Republican.

Later, as a West Side booster and engineer, Viele's interests diverged from those of Tweed and his Tammany cronies, who remained fixated on Upper East Side development and controlling the power of the municipal purse.[7] The West Point graduate spent the majority of his career seeking to make New York a world-class city that could rival London, Paris, or Berlin.[8] Unlike those municipally appointed developers beholden to Tammany and the "Brooklyn Ring," Viele pursued the construction of Central and Prospect Parks, the development of the West Side, and the improvement of New York's sanitation without automatic acquiescence to the directives of the party machine leadership. His first intentions frequently were to make New York a better place to live. A close second was Viele's concern for his personal legacy and reputation, even if that meant following the whims of Tammany leadership.

Unlike those earlier West Point men described in chapter 4 who had dedicated much of their energy to establishing the professionalization of civil engineers in New York, Viele used his talents and influence to advance the agenda of the West Side Association. Another subtle difference between Viele and West Point engineers who supported the American Society of Civil Engineers was that he leveraged his membership in New York's social clubs to advance his ideas. His efforts to promote himself were more transparent than those of the earlier West Point men.[9] In the last two decades of his life, the old general and city booster devoted his energies to his alma mater, speaking to cadets, serving on the Board of Visitors, and enlarging the West Point Cemetery (which would grow to include his 31-foot-high pyramid-shaped mausoleum).[10] In spite of his penchant for pursuing projects that tended to be self-serving, Viele's career in New York reflected his engineering-focused education at West Point, his army experiences in the Mexican War, and his service in the Civil War. Additionally, his family status as a Knickerbocker of New York underwrote his expertise and experience. Like George S. Greene and William Sidell, Viele was a secondary yet consequential player who shaped the course of New York's rise. Though Viele never achieved the power of Tweed, the fame of Olmsted, or the wealth of the moneyed elite in the city, like the other Military Academy graduates, he determined the course of debate over the direction and shape of the city. Through his speaking engagements, his written reports and articles, and his business pursuits, he encouraged his city peers to make New York a leading world metropolis. What is even more important, his actions were a harbinger of the reforms made by the Progressives, who emerged in the United States at the end of the nineteenth century.

In his 1998 book, *Other Leaders, Other Heroes,* James Endler notes that while serving as park commissioner, Viele "brought the Olmsted-Vaux design into being, personally serving as the chief engineer for the development and construction of Central Park."[11] Endler's assertion is incorrect. Throughout his career as an engineer and park designer, Viele was at odds with Calvert Vaux and Frederick Law Olmsted. Their rivalry began in earnest when the board of park commissioners created by the Republican New York State Legislature dismissed Viele as engineer-in-chief and promoted Frederick Law Olmsted to architect-in-chief of the park in May 1858. Viele and Olmsted would spend more than four decades challenging each other for the title of designer of Central Park. From Olmsted's winning control over Brooklyn's Prospect Park in 1865 to Viele's securing a position on the Central Park board of commissioners in 1883, the two competed with each other.[12] Contrary to what Endler maintained, Central Park was a source of perpetual discord between Viele and Olmsted.

While there is no doubt as to the significance of Frederick Law Olmsted and Calvert Vaux in the design and construction of Central Park, Viele's tenure as engineer-in-chief of the park remains an important, yet seldom emphasized, chapter in the park's history. The head of park development and construction from 1856 to 1858, Viele applied the lessons of his West Point education and his army service on the Texas frontier to create a detailed topographic survey and drainage plan for the original 776-acre park property. More importantly, he provided the necessary topographic information and an initial design from which Olmsted and Vaux would create their design masterpiece called Greensward. While Viele's 1856 design for the park lacked the sophistication and artistic mastery that Greensward's possessed, it was a crucial first step in securing the location of Central Park at the center of Manhattan. Without Viele's surveying, preliminary drainage work, and initial construction of the site, Olmsted and Vaux might not have had such a large plot on which to develop their masterpiece of landscape architecture.

THE ORIGINS OF CENTRAL PARK

From the start, the creation of a public park in New York City was to be a source of political and social tension. The construction of Central Park redefined the expanding role of municipal government in the lives of wealthy, middle-class, and working-poor New Yorkers. Central Park's influence on Manhattan extended well beyond the 843-acre property of 1863, affecting everything "from city planning and real estate investment to conditions of public employment and the city's fiscal integrity." Publicly, the park's founders proclaimed the creation of a leisure space as a common good for all classes of New York society; yet in the privacy of real estate and business deal-making, the construction of the park primarily served the interests of New York's richest and most powerful citizens—"its gentlemen and ladies."[13]

In 1844, romantic poet and newspaper editor William Cullen Bryant was one of the first New Yorkers to call for a park to facilitate health and provide a pastoral outlet in the city.[14] Wealthy merchant Robert Minturn, at the behest of his wife in 1848, demanded a public park similar to those he and his wife had seen during a recent visit to Europe. As early as 1851, Whig mayor Ambrose Kingsland proposed a park to be built at Jones Wood, a sprawling 150-acre expanse along the East River. For more than five years, the city's public officials, genteel elite, and speculative merchants campaigned for a park that would be central to the city and benefit all classes of New Yorkers. By 1855, through judicial and legislative actions, the state legislature designated the 776-acre property that extended from 59th Street to 106th Street, between Fifth and Eighth Avenues, to become Central Park. The state also allocated $1.5 million to construct the park, making it a profitable political prize in the city. All the while, the city Democrats and state Republicans deliberated as to who would control the project and the board of commissioners for the park. In 1856, Democratic mayor Fernando Wood bypassed the state assembly and appointed himself and Joseph S. Taylor, his street commissioner, as commissioners of the new Central Park. A consortium from the city elite supported this evasion of state authority and agreed to be Wood's consultants on park affairs. Among them were historian George Bancroft, banker Stewart Brown, former state senator James E. Cooley, and novelist Washington Irving, whom they chose as chairman.[15] In 1856, the Democratic mayor and self-proclaimed commissioner of the park appointed the thirty-one-year-old Viele to be the engineer-in-chief of Central Park.[16]

VIELE AND THE POLITICS OF THE PARK

The new chief engineer may not have been the best-trained landscape architect for the job, but he was unquestionably a well-connected choice. Viele was the son of John L. Viele and Kathlyne Schuyler Knickerbocker of Waterford, New York. His father was a prominent New York state senator and attorney. His mother was a descendent of the Knickerbockers of Schaghticoke in Rensselaer County.[17] In July 1842, at the age of seventeen, Viele entered West Point. Academically he was mediocre at best, but he learned the craft of topographical drawing as a second-class cadet in 1846. He finished tenth in his drawing class of forty.[18] Robert Weir's drawing course was Viele's best at West Point.[19] Under Weir's instruction, Viele and his fellow classmates learned how to use charcoal, pen and ink, and watercolors to draw terrain features, architectural designs, maps, and people. Additionally, Weir taught the cadets mechanical drawing.[20] Apparently Viele did not receive any other formal training in drawing after he graduated from West Point in 1847. Given that so much of his civilian career can be chronicled through his maps and topographical drawings, Viele's collections of designs and drawings reveal as much about his ambition as they do about the drafting instruction of Professor Weir. Weir's students had to demonstrate a high level of drawing skill in order to pass his topographical drawing course. Not only did army operations rely upon the accuracy of an officer's drawings, so too did civil engineering projects. Viele used the topographical skills he had learned at West Point to advance both his military and civilian careers.

Viele served two years in the Mexican War as a new infantry lieutenant under Gen. Winfield Scott.[21] He continued his first tour of service after the

Mexican-American War, serving at Ringgold Barracks and Fort McIntosh on the Texas frontier, where he helped build the military road between Rio Grande city and Laredo, Texas.[22] Viele married the former Teresa Griffin, a New York socialite, in 1850. Griffin's grandfather had been a law partner of George Templeton Strong's father, and her father had been a clerk at one point in Strong's law office.[23] Though Egbert and Teresa had eight children, their marriage would end in a scandalous divorce in 1872.[24] In June 1853, Viele resigned his commission and moved back to New York to establish his own civil engineering office. By 1855, he had secured the position of state engineer of New Jersey, where he "made a geodetic study of the state," establishing a name for himself as an accomplished engineer and topographer in the New York area.[25] Needing a loyal man to build his park, Mayor Wood convinced him to leave New Jersey and take the park position. Engineer-in-Chief Viele reported to Fernando Wood's park commission of city gentlemen.[26]

Throughout 1856, Mayor Wood fought a fierce political battle against the reform-bent Democrats and Republicans in the state legislature. Wood, one of the first mayors to harness the power of patronage and act as a "boss" in Tammany Hall, had been using the municipal police force as an instrument of political favor. Central Park was another outlet to grant favors and build support for the Democratic machine developing at Tammany. In late 1856, the state removed Wood's oversight of the city police by creating the state-appointed Metropolitan Police Board.[27] The legislature furthered their reform of city Democrats in the spring of 1857 by finally appointing a state-controlled board of commissioners to oversee Central Park. Replacing Wood's park commission that April were four Democrats, six Republicans, and a Know-Nothing who soon switched to the Republican Party.[28] Even though this council of gentlemen was bipartisan, the Republicans controlled it. Soon, they removed Wood's power of patronage in the park, only keeping Viele as the engineer-in-chief so he could complete his topographic study. By July 1857, the state legislature had stripped Wood and the Tammany machine of their control over Central Park and the New York City police.[29] Despite the political turmoil between Albany and City Hall, Viele would make the most of his position.

Prior to the appointment of the state's board of commissioners, Viele used his topographical findings to construct his own design for the park. Wood's commissioners approved Viele's plan for the park in July 1856.[30] Viele's design, his topographic study, and his drainage plan were his most noteworthy contributions to the park's creation.

In 1855, Viele used the topographic study to draw up his "Plan of Drainage for the Grounds of the Central Park," an elaborate scheme to drain the numerous "stagnant deposits" scattered throughout the property.[31] He was often concerned with the drainage of "stagnant" water that readily became "a pestilential spot, where rank vegetation and miasmatic odors taint every breath of air."[32] A faithful proponent of sanitation engineering in urban development, Viele suggested that "drainage of the Central Park [would] necessitate the construction of such drains along the whole slope between it and the East River, and this section [would], as a natural consequence, become the healthiest portion of the city."[33] In the drainage plan, he drafted in "great detail" all of the "hills, streams, and large rocks," as well as settlements within the park boundaries.[34] Viele's survey of the park not only portrayed the mixture of rocky and swampy relief of the property, it also recorded the shanties and cabin-like farmhouses where an estimated 1,600 people lived. His park surveys remain some of the best records of Seneca Village, a community of African Americans and some Irish Americans who lived just west of the rectangular-shaped Croton Reservoir (now the Great Lawn). Unfortunately for the residents of Seneca Village, their community did not survive the clearing of the park's land.[35] The "Plan of Drainage for the Grounds of the Central Park" was critical to the successful elimination of undesirable topographic features in the park, unwanted residents included.[36]

Viele possessed no empathy for the park inhabitants, instead portraying them as defiant squatters who threatened his efforts to master the geography of the park property. To his mind, these residents were of a lesser, unruly class who did not understand "the English language." Clearly, these New Yorkers were the type of folks Viele hoped the park would eventually reform, and perhaps even benefit.[37]

Regarding his design for the park, Viele noted that the "modern style" of park taste was "based upon the maxim, that 'the greatest art is to conceal art'" through a mixture of the natural and the artificial. He intended to create a masterpiece rivaling the great Hyde Park and Kensington Gardens in London. Viele sought through his design to "seek to know the peculiar wants of all classes, and to endeavor to gratify them at every step with a due respect to the principles of art, and an economical expenditure of money." Within his design, Viele expected to incorporate the existing valleys, streams, hills, and rock outcroppings to create a park that all New Yorkers could use and admire. He specified several manmade features as well, including a circuit drive for afternoon carriage drives, separate pedestrian and equestrian trails, a military parade field, a cricket ground, and botanical spaces. Viele designed the new 100-acre reservoir, called Croton Lake, to be "irregular in shape" in order to accentuate the aesthetics of "The Circuit" drive.[38] As noted in chapter 4, this was the same Croton Reservoir that George S. Greene actually built for the Croton Aqueduct Department.

Described in Viele's only annual report to Mayor Wood, in January 1857, the design appeared to be economical and in accord with current natural landscape practices. It conformed to much of the site's topography. Using this design, Viele might even have been able to complete the project within the allocated budget of $1.5 million. But Republican board members were not going to let a Democratic appointee control the destiny of park funds, patronage, and the design of Central Park, especially with Wood's influence marginalized by the actions of the state legislature in that summer of 1857. Instead, the state board would hold a design competition and award the park project to the best design from the most qualified "landscape gardener."[39]

OLMSTED AND VAUX

The most qualified "landscape gardener" proved to be a duo of architect and journalist-turned-landscape designer—Calvert Vaux and Frederick Law Olmsted. The two aspiring landscape designers first met in 1851 at the nursery of Andrew Jackson Downing in Newburgh, New York. Downing was the leading architect of garden design in the United States and had brought the English-born Vaux to Newburgh to be his assistant.[40] If not for his tragic drowning in a Hudson River steamboat accident in July 1852, Downing, who had been a popular park proponent, might well have become the park's architect. Instead, his diminutive protégé, Calvert Vaux, received the honor (Vaux was four feet, eleven inches tall).[41] Through his former boss's Republican connections, Vaux criticized Viele's plan in the summer of 1857, and convinced the state commissioners to scrap Viele's design and hold a competition. Looking for a qualified partner, Vaux approached Olmsted, who had become park superintendent under Viele at the behest of the Republican commissioners. Vaux thought that Olmsted would be best suited to prepare a design since he had been working in the park and knew the terrain. Together they submitted the elaborate proposal number 33 in March 1858, and, on 28 April, the state board of park commissioners awarded Greensward first place and $2,000 prize money.[42]

The Greensward plan was much more detailed than Viele's 1856 proposal. Olmsted and Vaux included meticulous descriptions of the various number and types of trees to be planted as well as numerous before-and-after sketches to show how they proposed to change the topography.[43] As an accomplished architect, Vaux most likely designed all of the ornate bridges and other structures, while Olmsted was responsible for the landscape schemes and overall construction of the project.[44] What set Greensward apart from the other entries were the four transverse roads drawn into the design. Vaux and Olmsted sank the roads below the park plain, thus allowing the park to appear as a continuous two-and-a-half-mile-long landscape, while simultaneously permitting city traffic to flow across the three blocked avenues.[45] Having awarded the competition's prize to Vaux and Olmsted, the state commission could dismiss Viele. Thus, in May 1858, began the open rivalry between Viele and Olmsted.

Olmsted bided his time among the Democratic workforce, clearing the park's "low grounds [that] were steeped in the overflow and mush of pig sties, slaughter houses and bone boiling works."[46] He resented the inefficiency of the state-created park commission. To his dismay, the eleven-man board was an unfit collection of lawyers and merchants who were "unmanageable, unqualified & liable to permit any absurdity."[47] When faced with a revenue shortage created by the Panic of 1857, the commissioners pressed Olmsted to release seven hundred of the nearly five thousand workers in the park, an act that would undoubtedly undermine the remaining workers' trust in the superintendent. And yet, Olmsted knew that he had allies on the state board. He wrote to his father that "[there] is a good deal of row among the Commissioners, the difference being chiefly referable to a greater or lesser degree of confidence in Viele. Those who are jealous of Viele strive to advance me—withdraw responsibility from him & confer it on me."[48] Apparently Viele failed to see Olmsted as an imminent threat, as he did not object to Olmsted's submission of a design for the competition.[49] Three weeks after Vaux and Olmsted won the design competition, the board of commissioners of Central Park appointed Olmsted as "Architect-in-Chief of Central Park," a position that combined the duties of "Chief Engineer and Superintendent."[50]

Disagreement over the adoption of the Greensward plan and the ousting of Viele became a public controversy in New York newspapers. Richard Grant White wrote a partisan-toned editorial defending Olmsted and Vaux's plan in the *Morning Courier & New-York Enquirer,* praising both Olmsted and his plan on 31 May 1858. Also appearing that day in the *New York Herald* and the

New York Daily Tribune were editorials criticizing the Greensward plan and claiming that Olmsted's appointment as architect-in-chief was a political job. The *Herald* even claimed that Olmsted was just a "farmer" who copied the basic design of Greensward from Viele's plan.[51] No longer in a position of power, Viele and his Democratic supporters challenged the legitimacy of Olmsted and Vaux's plan and appointment whenever possible. A libel case arose in December 1859 when Viele sued the publisher of the *Evening Post* for attributing an anti-Olmsted cartoon to him.[52] Viele won the case and continued to seek further legal action to mitigate the damage to his name and reputation as an engineer and gentleman of the city.[53] Clearly, Viele's status depended on the perceptions of his expertise, and later, his increased stature as a Civil War veteran and former general.

By 1864, the Civil War and landscape projects out west had temporarily taken Olmsted away from Central Park. Olmsted was in California during the trial that had been scheduled to establish authorship of the Central Park design, leaving only Vaux to testify in defense of the Greensward plan. In court, Vaux stated that he and Olmsted originally drew up the plan as a submission for the 1857 design competition. According to Vaux, his and Olmsted's design was more aesthetically pleasing than Viele's because the Greensward plan used drives that went diagonally into the park and used larger lakes. Both the drives and lakes gave the illusion of a landscape much deeper and vaster than the actual acreage between Fifth and Eighth Avenues.[54] Additionally, he and Olmsted sank the four transverse roads to mask the artificial traffic from the natural landscape. Olmsted later noted that the effect of the sunken roads was to give his "plan one stretch of unbroken view across turf from near the South drive near the Cricket Ground . . . to the North end of the Green," a view almost half-a-mile long.[55] Viele intended to keep his roads on level grade with the park terrain in order to provide maximum enjoyment for those passing through the park. Even though the court ruled that the City of New York had to pay Viele his salary and compensate him for his plan, topographic study, and legal expenses, the issue of authorship for Central Park's design would remain open to debate.[56]

During the trial, Vaux testified that $2,000 was an inadequate amount to be paid for the work done on his and Olmsted's Greensward design. Fearing that his testimony may have inadvertently helped Viele win the case and the substantial award, Vaux sought to challenge Viele before the gentlemen of the Century, the city men's club to which both men belonged. Without a proper forum, Vaux thought that his and Olmsted's design would be subject to public doubt as long as Viele's alleged authorship remained popular in the New York press. Fellow club member Henry Bellows convinced Vaux to drop the challenge for a confrontation with Viele in the club, since such a debate could have been unfavorable for the club's image. Instead, Olmsted's ever-growing reputation and numerous park creations over time would lead public opinion to celebrate Olmsted and Vaux as the artists behind the much-lauded design that became Central Park.[57]

In the years following the trial, there were occasional attempts to credit Viele as the designer and builder of Central Park. James Grant Wilson's four-volume *Memorial History of the City of New-York and the Hudson River Valley*, published in 1892, has a brief chapter on topography and parks written by Egbert Viele. Viele's chapter makes no mention of Vaux or Olmsted. Instead, he writes that after adopting "the design of Egbert L. Viele," the board assigned him "the duty of converting this cheerless waste into a scene of rural beauty in accordance with his design."[58] *Rider's New York City*, a 1916 guidebook, attributes the design of Central Park to Viele. Fremont Rider wrote that Central Park was "designed by Lieut. (later General) Egbert L. Viele, assisted by Olmsted and Vaux, landscape gardeners."[59] In 1967, Henry Reed and Sophia Duckworth attempted to reconcile the contributions of Viele, Olmsted, and Vaux to the park's design in their guidebook and history of Central Park, but they gave almost all of the credit to Vaux and Olmsted. Roy Rosenzweig and Elizabeth Blackmar's *The Park and the People* (1992) provides the most thorough and comprehensive history of Central Park. While they give detailed attention to the events between 1855 and 1864, Olmsted and Vaux's contributions tend to overwhelm the impact of Viele's work.[60]

Egbert Viele's contributions to the design were not nearly as beautiful, ornate, and extensive as the work of Olmsted and Vaux. Nor was Viele an innovative landscape architect who changed the way the world created public landscape gardens. General Viele's significance to the creation of Central Park was that he was the first to develop and envision on paper the potential for the swampy and rocky swath of land in the center of Manhattan. As engineer-in-chief of Central Park, he gave future park builders a comprehensive topographical study on which to base their contributions to the park. He created the more natural-looking round shape of the new reservoir. He gave Olmsted and Vaux a plan to continue the drainage of the swamps and make the area a magnificent public space. He began the construction of Central Park, and over the remainder of his life, Viele would argue to preserve the park's integrity and purpose in the city.[61]

SANITATION REFORM AND THE WATER MAP

Several months after the Central Park commissioners replaced Viele with Olmsted (17 May 1858), the West Point graduate continued to advocate for improved sanitation and drainage in New York. Presenting his assessment of the city's "sanitary affairs ... and their remedy" before the state senate Sanitary Committee, Viele recommended that the city not ignore the natural topography of Manhattan Island when building and improving the streets. Viele said "the health of a particular locality depended on its topography and not its latitude," and he presented "charts showing the topography of the island, and called attention to the evil of filling up running streams and ponds." In Viele's mind, there was a cause-and-effect between not allowing the land to drain naturally, and having the pooling and collecting of waters lead to disease and sickness, especially in neighborhoods built over old streams and springs. During that same presentation, he recommended raising the grade of streets in lower Manhattan and warned that "tenement houses should not be built as to exclude the sunlight."[62] This presentation was typical of the many forums Viele used to push "sanitary reform" before and after the Civil War.

Viele was not the only expert campaigning for sanitary reform in the Victorian era. Beginning with Edwin Chadwick in England and physician John H. Griscom in New York during the 1840s, "sanitary reform rested squarely upon an empirically grounded explanation of infectious disease." Known by medical historians as "filth theory," this view of the cause and effect of infectious diseases incorporated a wide range of conjectures and ideas that associated "putrefactive odors" from sewage and waste with the onset of yellow fever, cholera, typhoid, and diphtheria.[63] Accepted as conventional wisdom till the end of the century, filth theory contended that disease did not transmit from person to person, but rather originated from "putrefying organic wastes." Also implicated by sanitary reformers were areas of stagnant or standing water, wet ground, stale air, and places of little or no access to sunlight. Many of the diseases treated by sanitary reform were intestinal and thus were seen as abated by clean water and waste infrastructure.[64] Men like Chadwick, Griscom, and Viele actively sought ways to drain sewage and waste away from private side lots, built-up residential areas, and the disease-prone "impoverished districts" of a city. Given his experience draining the swamps of the Central Park property, and his appearances before the city government of New York, Viele readily subscribed to filth theory and the remedies of sanitary reform.[65]

According to Jon Peterson, sanitary reform influenced urban planning and development in three ways. Water-carriage sewerage, sanitary survey planning, and townsite consciousness not only stemmed the ills of "fetid waters" but also created alternatives to "piecemeal urban growth" in the nineteenth century. Chadwick was the pioneer in water-carriage sewerage. He mastered the technique of running water through egg-shaped sewer lines in London to draw wastewater away from cesspools and "choked up ditch drains."[66] John Griscom brought Chadwick's water-carriage sewerage to New York, first advocating it in 1842. In the United States, sewer construction resulted more from the need to improve areas for development than from the creation of and adherence to any single master plan. When Lemuel Shattuck conducted a census in Boston in 1845, he became alarmed at the appalling living conditions of poor immigrants and cautioned that all sources of filth threatened the city and needed to be removed.[67] Well through the Civil War, solid waste removal in American cities was piecemeal at best, left to pigs and other animals to scavenge off the streets, confounding efforts to carriage or dispose of water waste.[68] Throughout the 1850s, sanitary reformers championed centralized action under the mantra of public health. Cities, and especially New York, lacked the municipal power to emplace all of the proposed reforms and infrastructure planning called for by the new sanitation experts.

After the Civil War, though, at the behest of sanitary reformers, sanitary survey planning became the standard for American sewers and water systems. Engineers, physicians, and real estate speculators collaborated to build sewer systems that addressed the dangers of disease and poorly drained areas as a whole within city planning. The most famous case drawing attention to the problem was the yellow fever epidemic in Memphis, Tennessee, that killed 10 percent of the population living in the lower Mississippi River valley in 1879.[69] The National Board of Health, led by George Waring, worked to reconstruct Memphis from the bottom up, using sewers, roads, and building codes that mitigated the dangers of standing wastewater.[70]

However, sanitary survey planning could only really work in the event of a disastrous epidemic or calamity when reconstruction was generally accepted as the only option. More prevalent was the use of "townsite consciousness," or a general acceptance of "site arrangements" in urban areas that could facilitate better sanitation conditions.[71] Frederick Law Olmsted's and Viele's methods were models of the "townsite consciousness" approach to sanitation. Both men spent their professional lives as proponents of green spaces in urban areas, with Olmsted's efforts being much more celebrated and better remembered than Viele's. Again, Viele's interests competed with those of Olmsted. Granted, Olmsted appeared better qualified for sanitary reform, having served as head of the Sanitary Commission during the Civil War but, as described earlier, he clashed repeatedly with Viele over the purpose and design of Central Park and, later, Prospect Park in Brooklyn. Furthermore, Olmsted became a nationally known reformer in the course of creating urban parks and green spaces across the country.[72] Viele, on the other hand, confined his efforts to New York, where he also attempted to apply the other two methods of sanitary reform, water-carriage sewers and sanitary urban planning, to his projects. Thus, Viele's efforts in New York were representative of the three ways sanitary reform influenced urban development, especially in the second half of the nineteenth century.[73]

For nearly three years prior to the start of the Civil War, Viele was a staunch sanitary reformer in New York. In the spring of 1859, the National Quarantine and Sanitary Convention held its third annual meeting in the city. The convention appointed Viele to chair the Committee on Civic Cleanliness and the Economical Disposition of the Refuse of Cities. As chairman of the committee, which included five other sanitary reformers, Viele led the effort to create "plans for the disposition of Offal, Refuse, Street-cleanings, and Night soil of cities."[74] The committee wrote the following in its sixty-page report in 1860:

> If there were a city whose natural position was perfectly salubrious, and whose artificial constructions were all completed and based upon the principles of sanitary science, that city might be said to be in a normal hygienic condition that is, in a condition where the exercise of a proper degree of civic cleanliness would insure the health of the inhabitants. In order, therefore, to accomplish the full measure of sanitary reform in cities, it is necessary to bring them to this normal condition. To accomplish this, there are four leading subjects which demand attention in the order they are named, viz.: 1. Drainage. 2. Paving. 3. Supply of Water. 4. Sewage. When the municipality shall have completed these four necessary measures, and not till then, the responsibility for the health of the city rest upon the individual inhabi-

tants; and a compliance, on their part, with proper sanitary regulations, will undoubtedly secure an exemption from preventable diseases.[75]

While the scientific reasoning—most likely influenced by the "germ theory" recently advanced by Louis Pasteur—may not be wholly sound by today's standards, Viele's recommendations reflect a careful analysis of the environmental causes and effects of standing water and disease.[76] Just as novel as the remedies themselves was the expectation of the city to implement them. Viele and the committee saw civic cleanliness as a function and responsibility of the city or municipal government. Disease, epidemics, and their causes were too serious to be left to the forces of the real estate market and unconstrained capitalism. Moreover, better sanitation increased the value of real estate in a competitive urban market.[77]

Also highlighted in this report was a brief survey of how other cities achieved "cleanliness." The report described in detail how Paris maintained clean streets in the 1850s. According to the report, Paris employed immigrant sanitation workers who swept the streets daily and ran water down the drains at eight o'clock every morning without fail. This was Baron Georges-Eugène Haussmann's Paris, and points to the high regard that New York's boosters and reformers held for the French metropolis.[78] In this instance, similar forces of urban development were seen shaping social and political expectations of municipal powers on both sides of the Atlantic.

Viele used his military experience and title in the conduct of his sanitation campaign. In the process he furthered the perception that West Point graduates possessed the expertise expected of professional engineers and, in particular, topographical engineers. For example, in June 1859, he made his sanitary reform presentation before the New York Sanitary Association as "Lieut. Viele." During the presentation, he made his usual arguments for proper drainage of Manhattan's streams and waterways, but he also regaled the audience with his experiences curbing cholera in the Mexican-American War.[79] Viele used his military experience and knowledge to give legitimacy to his expertise in the sanitary reform campaign. In addition to relating tales of the Mexican-American War, Viele used a detailed topographical map during the presentation. With these types of presentations, he broadened civilian perceptions of the American military. Not only did army officers fight the nation's wars, but they also possessed a unique skill set honed by real-world experiences during their time in uniform. Ever the accomplished topographical draftsman, Viele continued to improve his map of Manhattan and its natural waterways. Eventually, he superimposed the Manhattan street grid over the terrain to show his audience the correlation of the insalubrious regions of the city with the natural water collection points on Manhattan Island.

The culmination of Viele's efforts was his 1864 Water Map and its accompanying report, *The Topography and Hydrology of New York.*[80] On the map, Viele depicted "the original watercourses, streams (underground and surface), meadows, marshes, ponds, ditches, canals and the shoreline before landfill expanded the city's boundaries."[81] A champion of defeating disease through the creation of proper sanitation systems, Viele hoped that the map would encourage government officials to end the practice of filling in vital ecological waterways in the name of city development.[82] Instead, the map became a starting point for building contractors, who used it "to determine whether their building sites are former riverbeds that could still flood foundations." According to Melvin Febesh, whose "company laid the foundation for the Citicorp building at Lexington Avenue and Fifty-Fourth Street," the Viele Water Map was accurate to a few feet, and helped prevent certain setbacks in the construction of the corporate high rise.[83] Incredibly, in the first decade of the twenty-first century, the map is still the starting point for any engineer or architect looking to do construction in Manhattan.[84]

But Viele's primary contribution to sanitary reform was his recommendations to improve the city. More as a point of fact than criticism, he noted that John Randel's grid design of New York streets and blocks made "no reference whatever . . . to the topography of the island." Consequently, the street grading created high embankments, especially in the upper sections of Manhattan, and poorly constructed culverts that failed to divert the existing streams. Typical of sanitary reformers of his day, Viele briefly recounted a history of plagues and disease in European cities as a warning to Gotham before proceeding with his recommendations. For the lower part of the city, the West Pointer admonished the city "to widen the narrow streets, and to raise the grade where streets pass through the original depression of the surface." He added, "Narrow streets, under any circumstances, are a curse to a city. They are too generally the abodes of vice and crime."[85] Here, the sanitary reformer became a social reformer, well before the Progressive era. By coming up with a physical way to change the city and address environmental problems created by stagnant water, Viele thought that he was also correcting, or eliminating, causes of bad social behavior. Even though his recommendations reflected the newer emphasis on public health over public behavior to stem disease vectors, elements of social admonishment and judgment still inform his reform rhetoric. During the cholera epidemic of 1832, Jacksonian society blamed the disease outbreak on the "sins of Americans," but just three decades later, Viele and his fellow sanitary reformers saw the real culprit as the physical filth of the regions where cholera appeared.[86] Instead of bad behavior causing the disease, Viele argued that bad sanitary conditions caused the disease, and the disease, in turn, led to bad behavior. So, by fighting disease with improved drainage and public works, Viele thought the city government could also curb unacceptable social behavior and improve the lives of the undesirable classes. In this sense, Viele was a precursor to the Progressives who emerged in the last two decades of the nineteenth century. But it would be his Civil War career, as well as sanitation practices used during the Civil War, that would prove to give Viele greater authority to proclaim the benefits of sanitary reform and to advance his vision for building a modern Gotham.[87]

VIELE'S CIVIL WAR: NORFOLK AND NEW YORK

When the Confederate cannons of Fort Johnson fired on U.S. Army troops at Fort Sumter on 12 April 1861, Viele, like many of his West Point alumni, re-entered uniformed service to combat the southern rebellion. Viele joined the Seventh New York Militia as a "captain of engineers" and was part of the defense of Washington, DC, during the first months of the war.[88] As an officer in the Seventh Regiment, Captain Viele published his *Handbook for Active Service* to aid his fellow citizen-soldiers from New York in mobilization and training to fight the South (and apparently his handbook inadvertently aided Confederate volunteers in Richmond, Virginia).[89] Following a letter campaign from his first wife, Teresa, and Viele's own lobbying efforts, he eventually secured an appointment as brigadier general in the U.S. Volunteers in late August 1861.[90] From 1861 through October 1863, Viele campaigned in Virginia, South Carolina, and Georgia, spending the preponderance of his time as the military governor of Norfolk, Virginia.[91] While his Civil War service was unremarkable, especially when compared with the service of his more famous West Point peers, Viele and his fellow volunteers were a source of pride for New Yorkers.[92]

But their support for Lincoln and the Republican Party's radical agenda was far from unanimous, particularly as the city's political forces competed for power throughout the Civil War. The Democrats split into "War" and "Peace" Democrats. War Democrats were the Tammany men, including Tweed, and businessmen who supported fighting for a Union victory, but opposed Republican partisan policies and Lincoln's curtailment of civil rights. Peace Democrats, based in Fernando Wood's Mozart Hall, favored restoring the antebellum Union and keeping slavery intact, and opposed emancipation. More extreme Peace Democrats, who backed Gen. John McClernand's intention to create "an independent 'Northwest Confederacy' of Midwestern states," advocated making peace "without reunion" as a last resort to end the war.[93] When the war dragged on, the Peace Democrats, led by Fernando Wood and his brother Benjamin, campaigned for peace, calling for a shift to the antebellum status quo.[94] Some New York Democrats, many of them businessmen, turned to the Republican Party in the patriotic fervor created by the attack on Fort Sumter, and favored defeating the Confederacy. To be certain, the New York Republicans, a mix of middle- and upper-class Protestants and reformers, emerged as the national leaders who came from New York during the war. Frederick Law Olmsted, as head of the Sanitary Commission, was just such an example. However, the Republican Party still remained the political minority in a metropolis of immigrant Irishmen, along with working-class and native-born Protestants.[95]

During Viele's service as a volunteer in the Seventh Regiment and later as a brigadier general, his support for the Union most resembled that of the elite Democratic businessmen and War Democrats. Even though his appointment as engineer-in-chief of Central Park had come from Fernando Wood in the 1850s, Viele's patriotism, sense of national duty, and loyalty to the nation all led him to embrace the war effort. By his own admission, the highlight of the Civil War for him was his accompaniment of Abraham Lincoln during the president's trip to the Peninsula battlefield in Norfolk, Virginia, in May 1862. In two laudatory articles written well after the war, he gushed about Lincoln, almost to the point that one would think Viele was a Republican.[96] Recalling his first meeting with the president, he wrote that Lincoln was "kind, genial, thoughtful, tender-hearted, magnanimous. . . . It was difficult to know him without knowing him intimately, for he was guileless and single-hearted as a child."[97] Two decades later, Viele's effusive praise of Lincoln continued. He boasted:

From that time until Mr. Lincoln's death I enjoyed the very closest intimacy with him. On one occasion he invited me to accompany him, the Secretary of War and the Secretary of the Treasury in a revenue cutter from Washington to Fortress Monroe. There was a small cabin in the boat divided by four partitions. During the period of eight or ten days we were together we never lost sight of each other. During the trip we were constantly engaged in conversation and discussion about war matters, much of the time being occupied in listening to Mr. Lincoln's wonderful fund of reminiscence and anecdote.[98]

Here, Viele's agenda is readily apparent: to improve his own stature through his association with a martyred Lincoln, especially as Lincoln's legacy increased in the three decades that followed the war. There were others who celebrated the service of the general, including New York reporter Richard Henry Savage, but Viele remained his own best publicist.[99] Despite Viele's embellishment in these brief memoirs, they reveal how a Democrat from New York City, as well as a West Point graduate, negotiated the politics of the time.

Volunteering to fight was just the first element of Viele's duty as a responsible Academy graduate and veteran of the Mexican War. Once given the mantle of command, in Viele's case as military governor of Norfolk, he had to follow and enforce the policy of Lincoln's administration no matter how unpopular or controversial that policy was. For example, with the Emancipation Proclamation, General Viele greeted "a procession of five thousand Negroes in Norfolk" who had come to wish him "a happy New Year, and congratulate themselves on the fact that they had the rights of freedom."[100] Knowing that the order did not apply to the occupied states, Viele nonetheless let Norfolk blacks think they were free. This may have been a case of Viele being more practical than ideological, especially since he and his family were living in the occupied southern city. His son, Egbert L. Viele Jr., was born in Norfolk on 23 April 1863.[101] As a white family with young children living in Norfolk during the war, the Vieles had to be sensitive to the changing relationships between slaves and freed blacks in Virginia. Despite reports of freed blacks making violent attacks on white Union troops elsewhere in Virginia, the Vieles had to maintain positive relations with blacks in Norfolk.[102] If Viele had remained in New York as a Mozart Hall Democrat aligned with Fernando Wood, he might not have been as accommodating to the idea of emancipation. Although Viele's position on freedom for the slaves may have been unpopular with New York Democrats, his actions in Virginia received positive coverage in the New York press.

In his capacity as military governor, Viele pursued a course to make Norfolk a functioning municipality, loyal to the republic and capable of providing basic public services. On Washington's birthday, he delivered an extemporaneous speech in honor of the first president and reorganized "the old Norfolk Fire Department," requesting new hoses and equipment.[103] By all accounts, Brigadier General Viele made his fellow New Yorkers proud as he led the disloyal and misguided southerners of Norfolk to enlightened reform modeled on the example of New York City.[104]

But Viele's commitment to the Union cause was not without limit. In October 1863, the army transferred the military governor of Norfolk to Ohio to supervise the draft there. By 20 October, Viele resigned his commission and returned to New York City to resume his engineering career and business.[105] Apparently, the New York West Pointer thought he had done enough to save the Union and needed "to attend to long neglected private interests."[106] Within three months of his homecoming, the directors of the Susquehanna and Wyoming Valley Railroad and Coal Company elected Viele as president of the company's board of directors, an opportunity that failed to bear fruit.[107]

The New York that Viele entered in the winter of 1863–1864 had changed during the tumult of the war and the draft riots of July 1863. Specifically, Tweed's Tammany had become the center of Democratic power in the city, as the Democrats had effectively reduced New York's quota for the Federal draft. Lincoln deftly avoided further provocations in New York by making a Democrat, Gen. John A. Dix, commander of the Department of the East after the riots ended.[108] By the fall of 1863, when Democrats August Belmont and Gov. Horatio Seymour started to rally support for George McClellan's nomination for the national ticket, Wall Street already appeared to be business as usual.[109] Democrats and Republicans alternately declared unity and blamed each other for the war's injustices manifest in the city. New York in 1864 was a microcosm of the intense maelstrom that Union politics had become during the war.[110] Unlike General McClellan, who became a public leader of the Peace Democrats' opposition to Lincoln's Republican Party, Viele spent the remainder of the war pursuing personal and business interests. He focused on constructing "country estates" at Rockaway and then Ashford Hill, two properties along the Hudson River in the area that would become the Upper West Side.[111] For the next two decades, Viele was one of the most active boosters of the West Side.

The West End Association formed in 1866 to "protect and advance" the interests of Upper West Side residents. Like-minded businessmen and land developers wanted the association to "systemize" a "haphazard approach to urban development." Like the other "boosters" (and sanitation advocates) in Victorian New York, the West Side men used Haussmann's Paris as their model and ideal. The *Real Estate Record and Builder's Guide* suggested, "[Despotic] governments are generally bad governments, but when one hears of the marvels Napoleon has accomplished in Paris, in the way of street improvements, it makes us wish that he, or someone like him, could be Emperor of New York for about ten years."[112] Under Emperor Napoleon III (Louis Napoleon), Haussmann had absolute authority to cut broad, tree-lined boulevards through Paris's ancient neighborhoods and narrow streets in the name of making Paris more beautiful. Additionally, Napoleon III and Haussmann intended "to improve public health and reduce crime, improve the flow of traffic and commerce, provide better sanitation with a vast new sewer system, improve the city's water supply, and provide more open space and clean air."[113] In just over two decades, Haussmann transformed the French capital into a modern marvel for all nineteenth-century urban developers to envy, including those in New York.

Although seen by many New Yorkers as inaccessible and too rugged for development, the Upper West Side with its "elevated plateau afforded magnificent views of the Hudson to the west and the splendid new Central Park to the east, and river breezes provided a salubrious climate."[114] Early on, the West Side Association sought ways to develop the West Side in the postwar era. Members proposed building a rapid transit underground railroad similar to the new underground being celebrated in London.[115] By 1868, the West Side Association was an example of how landowning New Yorkers could collectively advance their interests in public. Case in point: James R. Taylor and Gen. E. M. Barnum led the newly formed East River Improvement Association to lobby for construction of new wharves and piers on the East River, as well as the removal of dangerous rocks in Hell Gate.[116] Even though owners on the West Side were the first to form an association, they did not experience an immediate reward for their efforts. Tammany politics and geography favored the East Side. With Tweed's reemergence in 1869–1870, the Tammany men directed their real estate treasure and influence to the neighborhoods east of Central

Park. A leading member of the West Side Association and a Democrat, William Martin initially thought Tweed's influence in the park environs would propel West Side growth. Developing the East Side cost less in labor and the terrain was flatter, making it more suitable for grading and building than on the West Side.[117] Between 1864 and 1878, the development of the West Side languished in political intrigue, corruption, and nature's obstacles of stone and water; nevertheless, the West Side Association remained persistent in its goals for the next three decades.[118]

The West Siders jealously followed the rise of the East Side into one of New York's most desirable and fashionable areas. While expensive brownstones dominated building construction east of Central Park, property owners on the west settled for collecting rents from lower-class tenants who lived in shanties, cheap flats, and stalls. Larger landowners were concerned that the lower-class structures would discourage serious development.[119] William Martin led the West Side Association effort to convince the state legislature to grant the Central Park Commission authority to control the planning for the space above 155th Street, which was not covered in the 1811 grid map. The legislature granted this right and more to Andrew Haswell Green's park commission, eventually giving him jurisdiction over the Bronx—although Westchester landowners checked the Central Park Commission's expansion beyond the Bronx. But the Central Park Commission had become "the nation's first de facto planning agency," and was quickly followed by the Prospect Park Commission in Brooklyn under James Stranahan.[120]

Part of the development of the Upper West Side included a plan for "a second scenic boulevard, the much-debated 'Riverside-avenue.'" William Martin first proposed the avenue in 1865. Martin recommended changing the proposed straight-line grid of 12th Avenue with a combination scenic roadway and park that contoured to the shoreline of the Hudson River. Because initial plans conceived in the 1870s were impossible or not aesthetically pleasing, Martin and Viele designed their own. However, as in 1858 with Central Park, and in Brooklyn with Prospect Park in 1873, the city adopted a plan designed by Olmsted. The construction of what became known as Riverside Drive began in 1877 but was severely hampered by to inconsistent rock formations and insufficient landfill. Even so, architects and developers continued planning and building houses along the new avenue and in the Upper West Side neighborhood.

About the time the West Side Association gained traction at the end of the 1870s, Viele, now commonly addressed as "General Viele," became an integral member of the association. In 1872 he had built his "red brick, ivy-covered suburban villa" that occupied "a half-block of well-planted lawn on the southeast corner of 88th," and had become a fixture of the "West End Plateau" (as he referred to that part of Manhattan).[121] During the spring of 1879, the West Side Association reorganized under a new charter, and began to meet regularly at the Fifth Avenue Hotel. Under this new charter, the association came to depend on Viele's expertise in sanitation, park design, rapid transit, and the construction of Riverside Drive. Viele chaired subcommittees on "Parks and Public Places" and drainage. In the West Side Association, Viele found a vehicle to advance both his sanitary reform agenda and the value of his West Side property. Invoking the authority of the association, he requested that the Board of Health of the City of New York "make a special examination" of "blocks between 59th and 110th Streets, the Central Park and the Hudson River," and report the results to the city and the West Side Association. Citing the "large number of instances where the most palpable and disgusting violation of sanitary laws are permitted to manifest injury of the public Health," Viele sought to compel the city to clean up the shanties and areas of squalor in the West Side.

The West Side Association recognized that it was not a part of the municipal government, but it did have the persuasive power to lobby the city government to act in its interests. Unlike Tammany or other party machines, the West Side Association was a state-sanctioned organization that sought to act with some sense of transparency and use the power of its members' collective prestige to compel government action. As a corporate group, it had an effect on the city. A month after submitting the association's request for action, Viele reported to the membership that the municipal men had drained Manhattan Square (the current location of the Museum of Natural History), demonstrating evidence of some progress. Even so, there was still much to do to improve the West Side. High on the agenda was getting a means of rapid transit to connect the Upper West Side with downtown, completing Riverside Drive, and hosting the proposed World's Fair of 1883—which failed to come to fruition. Over the course of their gatherings, the men of the West Side Association remained fixated on these priorities.

Like other property owners, Viele found the West Side Association to be a means to cultivate expertise and support for developing New York City west of Central Park. According to its articles of incorporation, the West Side Association had four objectives:

1. To care for, protect and promote the proprietary interests of owners of real estate situated within the boundaries of Fifty-Ninth Street, Eighth Avenue, One Hundred and twenty-fifth Street, Manhattan Street, and the Hudson River in the City of New York, and to attend to matters of public concern affecting those interests.

2. To build, found and maintain within said boundaries for the uses of the Society an Association hall, with a library and reading room connected therewith, and to acquire, hold and own such real estate within said boundaries, and personal property as be necessary or proper, and which said Society may lawfully hold for the purposes of its organization.

3. To secure the economical, proper and efficient administration of the Government of the City of New York; to manage the property and revenues of the Society; to provide for the increase, limit and conditions of its membership; to conduct all necessary and proper business relating to its affairs; to print, publish and circulate books, documents and papers; to procure the passage of laws and ordinances necessary or appropriate to carry out the object of the Society, and

[4.] To mutually benefit the members of the Society by promoting such object.[122]

Similar to the American Society of Civil Engineers founded in the 1850s, the West Side Association hoped to increase its status and identity in the pursuit of a permanent building to hold meetings and house a library. Unlike the emerging professional organizations that had been dedicated to specific occupations in the second half of the nineteenth century, the West Side Association focused on public improvements that benefited its members, perhaps at the expense of other New Yorkers. For example, in late 1879, Edward C. Clark told the West Side Association that developers needed to build with more character, and contracted Henry J. Hardenbergh to design and build the Dakota Apartments as a way to drive shanty dwellers out of the West Side neighborhoods.[123] Viele understood the value the association had in fostering connections with the elite powers of New York. He often used the association's meetings as a venue to present his topographical findings and sanitation views.[124]

Through his West Side Association connections, Viele secured more influential positions, such as presidency of Central Park's board of commissioners in early 1883.[125] Expressing his chagrin over Viele's appointment, Olmsted lamented: "It has for twenty-five years been his [Viele's] principal public business to mutilate and damn the park."[126] Part of Viele's motivation for securing the Central Park position was to improve access to the West Side through Central Park. He once noted that a visitor from London wished to visit the Museum of Art on the east side of the park, but then could not find a way across the park to visit the Museum of Natural History on the west side. Viele hoped to remedy this type of obstacle to West Side access.[127]

While individuals achieved considerable personal and financial gain through their actions in the association, the membership recognized that they needed to remain committed collectively to West Side improvement in order to sustain progress. Together, they hoped to transform the West Side into a new Fifth Avenue where the rich could build mansions and, in the process, realize a profit.[128] In effect, the West Side Association members aped the trappings of professional and social societies, but ultimately for their private gain.

Even though Viele and his neighbors had purchased the land when it was cheap, they still tried to set a conscious example of their ideal development in the city. Viele's house dominated the West End plateau, with the Hudson River and Palisades providing a spectacular vista.[129] Given the sparsely settled pattern of Upper West Side housing construction in the 1870s, it is easy to see how West Side promoters thought they could make the neighborhood a bucolic getaway for the rich elite who lived in more established neighborhoods farther downtown. In an article for *Harper's New Monthly Magazine,* Viele wrote, "[My] house faces to the south, and a broad veranda extends around three sides of it—the south, the west, and the east."[130] However, the West Side leaders and speculators had missed their mark by trying to lure the New York elite with promises of grandeur and spectacular vistas from the West End plateau. Between the 1880s and the end of the century, Gotham's rich could escape to Bar Harbor, Maine, or the Adirondacks, or Newport, Rhode Island, to attain what Viele and company were offering.

Riverside Drive was finally complete by 1880, but not without more controversy. Claiming they were unpaid, the contractors blocked access to the completed roadway. When they posted guards and obstacles on the cross streets, residents like Viele had to find their way home through the "surrounding shanties, at imminent risk of being bitten by the vicious dogs the shanty dwellers kept to harass the bailiffs."[131] With incidents such as these reported in the press, New York's most wealthy failed to find value or any advantage in building mansions on the West Side.[132] Nonetheless, the efforts of the West Side Association had some effect on real estate and land development west of Central Park: first, by supporting Riverside Drive development, and second, by advocating rapid mass transit to the Upper West Side.

MASS TRANSIT AND ELEVATED RAILROADS

As mentioned previously, New York's boosters realized that creating an efficient public transit system could improve city neighborhoods and advance their personal interests. General Viele was most active with the West Side Association between 1880 and 1883, but he had started first as a mass transit advocate for all of New York. Both the West Side Association minutes and the New York press document a steady record of Viele actively campaigning and lobbying for completion of Riverside Drive and Park, the construction of Morningside Park, and extension of the elevated railway through West Side neighborhoods.[133] Rapid mass transit challenged Victorian Gotham. Using street-level conveyances like a horse-drawn omnibus or steam engine–drawn tram car worsened the traffic problem. Elevated railways could free up street space, but were unsightly and expensive to run and maintain. Underground rail transit had been proven to work with London's Underground in the 1860s but presented dangerous health risks.[134] Regardless of the risks involved, gaining and maintaining control of public transportation conveyances was a lucrative prospect in American cities. Urban and planning historians have coined these mixtures of private and public ventures "growth machines." While that level of power eluded Viele, he and his fellow boosters fashioned a model for eventual monopolists to concentrate urban power at the turn of the century.[135] In post–Civil War New York, businessmen and developers, Viele included, attempted to build all three types of mass transit.

Robert A. Chesebrough, the inventor of Vaseline, patented a design for an elevated system of locomotion in 1868.[136] Chesebrough enlisted General Viele to endorse his patent through a letter Viele published in the *New York Herald*, and subsequently in a short book. Here was an example of Viele lending his name, backed by war exploits and engineering expertise, to support an invention by someone outside the engineering realm. Viele explained that "[this] invention proposes an elevated railroad, the track of which consists of a series of inclined planes, down which a car runs by its own gravity, elevating platforms being interposed to raise the car from the foot of one incline to the head of the next." Both men described a railway that used air-compressed piston platforms and elevated tracks to raise passenger cars that would then coast down to each station stop. The published schematic of the piston, railcar, and sloped rail lines resembled something more like an amusement ride than a plausible plan for mass transit. While the invention may have been more Jules Verne than reality, Viele and Chesebrough were attempting to apply science and engineering to solve the problem of moving New Yorkers from their uptown and Westchester County residences to their places of business in downtown Manhattan. They also recognized that property values in areas adjacent to rapid transit stations that connected with the city would inevitably increase.[137] Chesebrough's elevated railroad system failed to progress beyond theory and Viele's public endorsement, but it did encourage others to think about how to move the ever-growing population of Greater New York.

In 1872, Viele published a report advocating the construction of an underground railway. He collaborated with several prominent engineers who were members of the American Society of Civil Engineers, specifically, William McAlpine, Julius Adams, and D. C. McCallum.[138] The report culminated a year-long campaign to persuade the state legislature to authorize funding for the city to hire an engineer and construct the railway. Also part of the lobbying effort was a model located in Albany and detailed schematic drawings showing a cut-away view of the "Arcade." A key concern for Viele and his partners was the safety and health of passengers riding the steam-driven cars below the surface of the street. Using London as their main example, they noted that "pure air and ventilation" were the only ways to mitigate "deaths from foul air, miasma, and suffocation." Instead of using tightly constructed tunnels that ran with intermittent open track to vent steam exhaust, the plan for the Arcade proposed an underground railway the width of the entire street with sidewalk grates above to ventilate the trains running below. The proposal also noted the advantages of excavating from the surface downward instead of trying to dig tunnels as the builders in London had done. The rocky geology of Manhattan Island made excavation construction the better course of action.[139] Like several of his earlier forays into the development of the city, Viele's underground railroad failed to materialize under his leadership. City Hall's post-Tweed reform sensitivities led to the creation of the Rapid Transit Commission (RTC) in 1875. Dominated by businessmen appointed by the mayor, the commission opted for steam-powered elevated railways as the way ahead for Gotham's rapid transit solution, and they awarded the contract to the New York and Metropolitan Elevated Companies.

Still, Viele persisted in being part of developing plans for rapid transit, at least for the West Side. As noted already, the East Side developed faster and more lucratively than the West Side, and certainly the Manhattan Elevated Railway Company, which was created by the Rapid Transit Company, contributed to this disparity. By 1880, there were three elevated lines running the length of Manhattan, and two of them ran up and down the East Side of the island.[140] Viele and his fellow West Siders recognized the handicap dealt to their neighborhood by this inequity, and stepped up their efforts to get more elevated rail lines west of Central Park.[141] In the fall of 1881, Viele presented "a plan of the electrical railway as operated on the Place de Concorde, in Paris, during the late exhibition, which he believed would answer the purpose of giving rapid transit to the West Side." Noting that this electric railway was already working in Berlin, Viele intended to place this proposal, as well as other West Side–specific projects, before the state legislature that winter.[142] By the next spring, he had decided to position himself in a more official capacity, seeking a spot on the Department of Parks Commission in order to advance the interests of the West Side in Gotham's growth.[143] Sitting on the Parks Commission, Viele balanced the need for rapid transit with the desire to keep the city's green spaces untarnished.[144] If Viele could not control a rapid transit project, then he and the West Side Association would try to shape and influence those who did.

CONCLUSION

To advance his interests in Victorian Gotham, Egbert Viele relied upon three key experiences: his Knickerbocker upbringing, his days at West Point, and his military experience in two wars. First, Viele's family background and his distinction as a Knickerbocker provided him with an early basis and the status not only to secure an appointment to West Point, but also to place him in New York's elite at various points in his life.[145] As a young lieutenant, he was a prized bachelor among New York's antebellum elite. Teresa Griffin's father reportedly was only too eager to marry his daughter off to a West Point man and Knickerbocker to boot.[146] Second, at West Point, Viele mastered topographic drawing, a skill that made him invaluable to Fernando Wood and to New York's sanitary reformers. Third, as "General Viele," he could leverage the fame of fighting in the Mexican War and the Civil War to enhance his credentials and expertise in the growth of postbellum New York. Of these three experiences, his mastery of topography was his most enduring and tangible contribution to Gotham's nineteenth-century expansion. Viele's drawings of Central Park and his Water Map became the basis for further construction in Manhattan.

Viele's body of drawings also exhibits the expertise that New Yorkers came to expect of West Pointers. The ability to portray the Manhattan landscape in a way that could be used by speculators, developers, and politicians made Viele a valuable ally. Municipal Democrats could use his work as a foil for the state Republicans' use of Olmsted and Vaux when arguing the merits of park design. True, Viele gained leadership experience through his military service, but his topographical drawing skill set him apart in New York. Also, he leveraged his skills in this area along with his writing to be an early advocate of proper sewage and water sanitation. Once noticed by Fernando Wood, Viele increased his status and responsibilities, eventually becoming a leader in the West Side.

Just as enduring as his drawings, but more difficult to distinguish, was Viele's leadership in shaping the spirit of Victorian New York. As a member of the West Side Association, he was often the voice of its lobbying efforts as he traveled to Albany and published editorials in the papers. By pursuing his own interests and personal aggrandizement, he also advanced the cause and interests of his fellow West Side property owners. General Viele combined the roles of booster, speculator, and reformer in the Upper West Side.

What needs further exploration, however, is how Viele and his fellow boosters, speculators, and reformers influenced the dynamic political and social scene that emerged in New York after the Civil War. The contest between Viele and Frederick Law Olmsted is but one aspect of the struggles among political forces who vied to control New York and reform the Victorian world around them. Viele's power and importance diminished as Gotham transformed into the premier metropolis of the twentieth century. However, the central experiences of being at West Point and serving in the Civil War remained important catalysts for Viele and his fellow West Pointers working in New York after 1865.

6

TOWARD CONSOLIDATION
Bridges, Bosses, and Brooklyn

This park beats Central Park ten to one in trees. Its wealth of forest is almost en-
viable. I think we cannot match its softly undulating lawns. But we beat it in rock
and the boxes of landscape. Prospect Park's attempts at rock are pitiable, most
palpable piles of boulders. We beat it also in water and bridges and other like
structures. But it beats us in views and is a most lovely pleasance.

—GEORGE TEMPLETON STRONG,
18 July 1871, from *The Diary of George Templeton Strong,* Vol. 4

George Templeton Strong toured Brooklyn's Prospect Park in July 1871
and, inevitably, compared it to Central Park. At the time, Brooklyn
was separate from New York, if not an up-and-coming rival to the
Victorian Gotham flourishing in Manhattan. The forces of reform, the impulse
for order, and a nineteenth-century sense of technological progress had subor-
dinated Brooklyn to Greater New York in less than three decades. Separated by
the narrow but treacherous waters of the East River, Manhattan and Brooklyn
were the first- and third-largest American cities in the second half of the nineteenth century. Manhattan was New York, the "American city" brimming with
all the wealth, culture, and squalor that signified industrialized civilization.
Brooklyn, on the other hand, was more pastoral, more spacious, and less egre-
gious in its filth and foulness. Brooklynites could enjoy the benefits of living
on tree-lined streets or in working-class apartment flats without the exorbitant
costs of similar accommodations across the river in New York.[1] Just a ferry ride
away from the hustle and harried existence of Manhattan was a place to which

New Yorkers could escape, such as Strong did that summer's day in 1871. Conversely, the same ferries carried laborers from Brooklyn to Manhattan, where the majority of Brooklyn's working-age residents were employed in New York's shops and businesses. Charles Dickens noted that "Brooklyn [was] a kind of sleeping-place for New York."[2] Essentially, Brooklyn was one of the first bedroom communities in the United States. So it was both work and leisure that drew Brooklyn and Manhattan together and spurred New Yorkers on both sides of the East River to find ways to unite the surrounding boroughs under one great municipal government.

Shared values and ideas drew the boroughs together. The park movement begun by the fathers of Central Park in the 1850s was one example of an idea spreading from Manhattan to Brooklyn, specifically through the creation of Prospect Park. Brooklyn's idyllic pastoral gem, Prospect Park, like its Manhattan counterpart, had been another source of competition between Viele and Olmsted, but never seemed to surpass the prestige of Central Park or any other enterprise in New York.[3] Until Brooklyn and the other boroughs officially became part of New York City, it and its public places would always be viewed as second-rate. In 1898, control of the parks transferred to a single municipal entity during the consolidation of Greater New York. Central Park, Prospect Park, and other public spaces across the five boroughs were subject to the direction of the New York Parks Commission.[4] The merger of Manhattan, the Bronx, Queens, Brooklyn, and Staten Island was not an inevitable outcome of the political posturing that occurred during the 1890s. Engineers, political leaders, and reformers, all motivated by the prospects of a better, more efficient metropolis, led the metropolitan elite to unite the boroughs.[5]

Post–Civil War New York was divided politically and, quite literally, physically over proposals to combine the boroughs under one metropolitan government. Once engineers and builders removed or bypassed the physical impediments to New York's consolidation, the political obstacles became more negotiable and easier to overcome. Washington Roebling's great East River Bridge (the Brooklyn Bridge) connected Manhattan and Brooklyn in 1883. Removal of the treacherous Hell Gate from the East River in 1885 made river crossings and navigation safer and more routine. Opening in 1895, the Harlem River Ship Canal enabled New Yorkers to sail around the island of Manhattan as well as reach the Bronx via bridge, boat, or ferry.[6] As Gotham's electorate decided their political fate during the consolidation referendum of 1894, West Pointers, with their engineering peers, had already built much of the infrastructure that connected the city more permanently than any single referendum could have done.[7]

Because of the "manifold possibilities" and "intense" life offered by post–Civil War New York, the premier American metropolis appealed to many former Union Army officers, most notably, Generals Ulysses S. Grant and William Tecumseh Sherman.[8] Generals Viele and George S. Greene returned to the engineering businesses that they had left behind in the city before the war. General George McClellan toured Europe for almost three years before coming back to the city in 1868. General John Newton remained in uniform as a lieutenant colonel in the Corps of Engineers, supervising the "surveys and improvements of the waters around New York City" for twenty years, and then served as the city's commissioner of public works.[9] Henry Warner Slocum abhorred the Radical Reconstruction advocated by Republicans and reinvented himself as a Brooklyn Democrat in 1866.[10] These former West Pointers and war veterans were in New York during a period of remarkable transformation and transcendence made possible by the great national sacrifice during the Civil War. If antebellum New York had been the nation's emporium, postbellum Gotham became the first American metropolis, attaining status as a world city.[11] As America's "first city," New York epitomized modernity, replete with all that people and technology could create. In this light, nineteenth-century engineering was the acme of progress; witness the Empire City that rose literally from the terra firma of Manhattan. With this spirit of the times, the former West Pointers' postbellum relocation to New York and Brooklyn was transformative for both the city and these men. In New York, West Point alumni marshaled their experience and expertise along with their Military Academy and wartime relationships to create their place in the modern city that emerged during the postwar era.

SLOCUM AND THE GREAT BRIDGE

Nowhere was this transformation of city and individual more evident than in Henry Slocum's building of the Brooklyn Bridge.[12] For many New Yorkers, the name Slocum reminds them of the burning and sinking in 1904 of a steamship named in his honor. The incident killed some 1,021 passengers, many of them German immigrants on a church picnic excursion.[13] Prior to that tragedy, though, the name Slocum was connected to the Civil War hero and Brooklyn congressman. General Slocum, as he was known after the Civil War, gained national notoriety by holding the "fish-hook" on Culp's Hill during the Battle of Gettysburg. He also received acclaim for reinforcing Union forces at Chickamauga and leading Sherman's left wing in the march to the sea of 1864.[14]

When the war ended, President Andrew Johnson placed Slocum in command of the Military District of Mississippi. Stationed in Vicksburg, General Slocum intended to enforce the emancipation decree for the freedmen while also following President Johnson's orders to work with the provisional Mississippi governor, William L. Sharkey, in reconciling the "new" state government with the Union "as quickly as possible."[15] While he was military governor, Slocum wanted to be fair to southern whites and freed blacks, but he also wanted to limit the use of military force in making them accept one another as equals. Slocum did not believe that the Constitution allowed the military to enforce racial equality.[16] One author described Slocum's position as "naïve," bordering on equivocation. In Slocum's effort to maintain a moderate stance on Reconstruction, he angered radicals by not offering freedmen sufficient protection. He also angered President Johnson and the Democrats by interfering with Governor Sharkey's Mississippi militia.[17]

The coup de grâce for Slocum came in October 1865, when he spoke before a Democratic rally at Shakespeare Hall in Syracuse, New York. During the speech, his first as a Democratic candidate for New York's secretary of state, he told the crowd that southern whites would be "humane" and have a "kindly impulse" toward freedmen. The *New York Times* disparaged Slocum's new party allegiance to the Democrats, the party that "[endeavored] to balk the war," by commenting, "It is a pity that he should make such a sale of his laurels."[18] With his political future seriously curtailed in central New York, a hotbed of radical Republicanism, Slocum left Syracuse in the spring of 1866 and started anew by opening a law practice in Brooklyn.[19]

Slocum's background helps to explain the general's personal transformation in Brooklyn after the Civil War. Slocum was born and raised in the village of Delphi Falls in Onondaga County, New York. The sixth son of a village merchant, Slocum attended the Delphi Public School and later studied at the Cazenovia Seminary to become a schoolteacher. As a nineteen-year-old, he read biographies of Napoleon and stories about the Mexican War, which led him to apply to the United States Military Academy in the summer of 1848. His classmates in the class of 1852 included Philip Sheridan. Other notable West Point graduates present during Slocum's cadet days were John M. Schofield,

Gouverneur K. Warren, and Oliver O. Howard.[20] At West Point, Slocum studied drawing under Robert Weir, took D. H. Mahan's engineering and military science courses, and graduated seventh in a class of forty-seven. A seasoned teacher, Slocum readily assisted his fellow cadets with their studies.[21] Commissioned as an artillery officer, he served in the Seminole Wars from 1852 to 1853 and then finished out his service obligation in garrison at Fort Moultrie, South Carolina, between 1853 and 1856.[22] After resigning from the army in 1856, he became a lawyer in Syracuse, where he remained until the Civil War broke out in 1861.[23] In the prelude to the rebellion, Slocum became a cautious Republican, serving as representative in the New York state assembly and freely voicing his opinions on contentious issues, regardless of the party's position.[24] What appears to be consistent with Slocum is that from West Point, through his army duty at Fort Moultrie, and later as a lawyer and state representative, his political allegiances and expectations were predicated on his own beliefs and understandings more than the opinions of those over and around him. Moreover, when a policy or party plank ran counter to his judgment, Slocum thought that he was duty-bound if not obliged to communicate his disagreement.[25]

But as a general for the U.S. Army in the Civil War, Slocum tended to fall in line with guidance from his superiors. Slocum biographer Brian Melton characterizes General Slocum's Civil War service as loyalty to his commanders almost to a fault, imitating their best and worst traits. Subsequently, countless battle histories, including the *Official Records of the War of the Rebellion,* portray Slocum's actions on the battlefield as ancillary to the heroics and directives of his Union commanders.[26] During General McClellan's tenure as commander of the Army of the Potomac, Slocum, who was still a Republican, was one of his most loyal subordinates. For instance, Slocum readily adopted McClellan's "policy of conciliation toward Southern civilians" by calling for a formal inquiry into reports of Union forces pillaging the Virginia countryside in October 1861. When officers in Slocum's brigade revolted in protest against this lenient policy, Slocum, following McClellan's example, disciplined the dissenting leaders to reinforce his command over the brigade.[27] Throughout the autumn and winter of 1861, General McClellan repeatedly frustrated Lincoln by exaggerating Pinkerton's intelligence reports of Confederate strength and disposition and using those inflated reports as a reason not to engage the Confederate forces with his army.[28] By most accounts, although Slocum took an aggressive approach with his troops by taking to the offense over defensive actions, his battlefield fortunes remained tied to McClellan's hesitancy and deliberateness.

Moreover, Slocum's loyalty to McClellan lasted well beyond McClellan's relief from command in 1862, which later led to vehement discord between Slocum and General Joseph Hooker (USMA class of 1837).[29]

The major blemish in Slocum's Civil War record resulted from this bitter feud with Hooker. At Chancellorsville in May 1863, Hooker squandered an opportunity to defeat Lee's army, and during the Union retreat, Slocum's Twelfth Corps suffered substantial casualties, losing almost one-third of his command.[30] By the late fall of 1863, both men had lobbied President Lincoln to have the other removed from the field. The president left each in command of his respective corps in a less than ideal situation for waging war in Tennessee that winter.[31] Slocum's virulent antipathy toward Hooker undermined his effectiveness as a commander in the western theater. Only after General Sherman took command of the Union Army in the West did the New Yorker's performance and reputation rebound, specifically with the Union victory at Atlanta, whose mayor surrendered to him.[32] Just as he had mirrored the actions of McClellan earlier, Slocum followed Sherman's example during the march to the sea. Fortunately for Slocum, Sherman was aggressive, and so was the left wing under Slocum's command. Sherman's good press rubbed off on him enough to balance the ill effects from the feud with Hooker.[33]

Slocum's military legacy may have suffered from his proclivity to mirror the commanders over him, but his loyalty to his superiors, save Hooker, rehabilitated his political career after he switched political parties in 1865. General Sherman wrote him in March 1868: "As to politics, it is impossible for language to convey my detestation of them. I have seen Fear, Cowardice, Treachery, Villainy, in all its shapes contort and twist men's judgment and actions, but none of them like politics. . . . They have tried to rope me in more than once, but I have kept out and shall do so as long as I can; and then I hope I shall die before what little fame I have is lost and swept away."[34] Here, Sherman expresses empathy for his old comrade who had chosen a political life. Sherman and Slocum continued to correspond after the war, and whatever differences had emerged during the 1864 campaign to the sea were pushed aside and forgotten. Sherman demonstrated an unwavering loyalty to Slocum, openly supporting his political campaigns in 1865 and 1868.[35] Slocum's performance leading the Army of Georgia in Sherman's victorious campaign overshadowed any negative thoughts Sherman might have had about his former subordinate. The more time passed after the war, the more revered the men's wartime experiences became, and Slocum's war exploits and memory were no exception. Both men's legacies would be closely tied for the remainder of their lives, with Slocum chairing Sherman's funeral committee in 1891.[36]

Key to Slocum's postwar career was his ability to foster those army relationships, some of which started at West Point, and many more of which emerged during his Civil War service. Four of the general officers in the Peninsula Campaign of 1862, Generals McClellan, Newton, Porter, and Slocum, spent much of their postbellum careers in New York.[37] The significance of their shared wartime connections was not apparent until well after hostilities ended. For a case in point, the reputations of Slocum's pallbearers were indicative of the status that Slocum achieved in postbellum New York. Among the Civil War veterans who laid Slocum to rest in 1894 were two West Pointers, four Medal of Honor recipients, and a Republican U.S. attorney. While most of the pallbearers had accompanied Slocum at various battles in the war, Gettysburg chief among them, the postwar associations that Slocum cultivated through the Grand Army of the Republic and New York's political and legal communities deepened their ties beyond their common Civil War experience.[38] In life and in death, Slocum was among his peers, men who had proven themselves in war and, for good or ill, became well-known figures in postbellum New York.

For Slocum, though, his stature and reputation were very much in the balance when he moved to Brooklyn in 1866. His professional and political transformation began with two opportunistic circumstances: one, the Brooklyn Democrats had very few Civil War heroes to claim among their ranks and, two, the campaign to build the East River Bridge spearheaded by Boss McLaughlin's Democratic machine, the Brooklyn Ring, was forging ahead in the late 1860s.[39] When William C. Kingsley and James S. T. Stranahan invited General Slocum to purchase shares and sign on with the board of directors of the New York Bridge Company in 1869, Slocum embraced the opportunity. He bought five hundred shares and became an active board member.[40] Also in 1869, Slocum became Brooklyn's representative in the U.S. Congress. Serving in the Forty-first and Forty-second Congresses as a Democrat, he advocated federal support for the bridge.[41] Wary of the Brooklyn Ring, Slocum was careful to keep his seat by staying clear of the Democratic machine and remaining close to Kingsley, perhaps the most powerful supporter in Brooklyn of the bridge.[42]

In order to win support for John Roebling's wire suspension bridge design, the Bridge Company sponsored a tour for politicians and engineers to inspect Roebling's bridges at Pittsburgh, Cincinnati, and Niagara Falls in the summer of 1869. While the bridge entourage was traveling to Pittsburgh and

Cincinnati, Roebling, who had remained in Brooklyn, crushed his foot in what appeared to be a minor ferry accident and then died weeks later of a tetanus infection.[43] Slocum, who was among the bridge delegation, joined the chorus of eulogies, clearly playing up his Civil War record for the sad occasion. At Niagara Falls, Slocum reflected on Roebling's suspension bridge accomplishment by saying it was "the only thing for which he would be willing to forfeit his war record, and to have been the engineer of the Suspension Bridge he would have gladly dispensed with whatever honor he might have won during the war."[44] While this may have been a case of Slocum engaging in nineteenth-century hyperbole, his sentiments do reflect a certain deference to, if not religious reverence for, the "science of engineering" in Victorian America. Slocum may not have been an engineer but, to be a political leader in that age, he had to recognize and embrace the spirit of the day, a spirit embodied in the celebration of progress through engineering.

Louis Sullivan, the great architect and father of the modern skyscraper, reflecting upon coming of age in the mid-nineteenth century, noted that "[the] chief engineers became his heroes; they loomed above other men."[45] Roebling was just such a hero. Born and raised in Prussia, Roebling studied engineering in Berlin before emigrating to the United States in 1831. He possessed the expertise and ambition to guide the most challenging projects to completion over the span of his career.[46] Roebling's untimely death made him a "martyr" to the bridge and the East River Bridge a memorial to him. The *Brooklyn Daily Eagle* captured the essence of the moment: "He who loses his life from injuries received in the pursuit of science or of duty, in acquiring engineering information or carrying out engineering details, is truly and usefully a martyr . . . we look on the great project of the Brooklyn Bridge as being baptized and hallowed by the life blood of its distinguished and lamented author."[47] If Lincoln's martyrdom saved the Union at the end of the Civil War, then Roebling's death ensured that the dream of a great bridge over the East River would be fulfilled. Slocum's speech at Niagara Falls was another example of comparing civil engineering with battlefield heroism. The West Pointer, like the *Brooklyn Daily Eagle* (whose editor, Thomas Kinsella, was also on the board of the Bridge Company), was capitalizing on the public's deification of engineering to improve his political and business interests in completing Roebling's bridge.[48]

In this light, the Brooklyn Bridge was another monumental project that drew engineers to New York. Similar to the building of the Croton Aqueduct, the creation of Central Park, and the rise of the American Society of Civil Engineers, construction of the bridge over the East River attracted civilian- and West Point–trained engineers. Additionally, the U.S. Army Corps of Engineers, under Gen. A. A. Humphreys, had to approve any span across the busiest seaport in the United States.[49] The chief of engineers, Humphreys, an 1831 West Point graduate, was the approval authority for any project that could affect navigation on the nation's waterways, which included New York's harbor and rivers.[50] For the inspection of Roebling's existing structures in 1869, Humphreys assigned three army engineers—Gen. Horatio Wright, Gen. John Newton, and Maj. William Rice King—to accompany the entourage of engineers and politicians.[51] All three were Military Academy graduates and Civil War veterans: Wright graduated in 1841, Newton in 1842, and King in 1863.[52] Approval by the army engineers enabled the East River Bridge project and, specifically, Roebling's design for the bridge, to proceed in 1870.

EMILY WARREN ROEBLING

Other engineers who joined the effort included Julius W. Adams, William Jarvis McAlpine, James P. Kirkwood, and Benjamin Henry Latrobe, all among the founding members of the American Society of Civil Engineers in 1852.[53] As previously noted, Adams had attended West Point for one year in 1831. But Adams was not the most pivotal West Point connection in the construction of the East River Bridge. John Roebling's daughter-in-law, Emily Warren Roebling, was. Emily was the wife of Washington Roebling, the son who took over the family business after John Roebling's death in 1869. As has been well chronicled in the twentieth century, Emily proved to be the most important catalyst for completing the bridge.[54] Her husband, Washington Roebling, was an 1857 graduate of the Rensselaer Polytechnic Institute. He served in the Civil War under Gen. Gouverneur Kemble (G. K.) Warren, Emily's older brother.[55] General Warren, like Henry Slocum, was a West Point graduate, as well as another "hero" from the Battle of Gettysburg.[56] Thus, the sister of a Military Academy graduate arguably had the greatest role in ensuring completion of the Brooklyn Bridge.[57]

G. K. and Emily Warren were two of twelve children born to the Warren family of Cold Spring, New York (only six lived past their childhood). Located in Putnam County, Cold Spring was just across the Hudson River from West Point and was the home of the West Point Foundry, the ordnance factory that produced the Parrott artillery gun.[58] Named for Cold Spring's most famous resident, foundry owner and politician Gouverneur Kemble, G. K. Warren was thirteen years older than Emily. With Kemble's encouragement and support, G. K. was destined to attend the Military Academy in 1846. By the time he graduated in 1850 and began his career as a topographical engineer in the army, young Emily was accustomed to the presence of foundry cannon and uniformed men in and around her Hudson Valley town.[59] When the Warrens' father died in 1859, G. K. Warren was assigned to West Point as a math professor. One of the advantages to being a Military Academy professor was his proximity to his family's home, where he could look after his younger siblings, including Emily.[60] With his army income, G. K. footed the tuition for his sister to attend the Georgetown Visitation Convent Academy in Washington, DC. Emily attended Visitation for two years, studying a rigorous curriculum that included history, geography, rhetoric, algebra, geometry, and geology.[61] G. K. ensured that Emily and the other Warren siblings would be well prepared for advancement as adults.[62] The outbreak of the Civil War abbreviated G. K.'s tour in the Math Department, and he served in the Army of the Potomac for the duration of the war. For both G. K. and Emily, the war changed their lives.

Like other West Point graduates fighting for the Union, G. K. Warren quickly rose through the ranks from lieutenant to major general. By the Battle of Chancellorsville in the late spring of 1863, General Warren was the chief engineer of the Army of the Potomac, whose duties included leading the staff of topographers.[63] At Chancellorsville, Washington Roebling was assigned to Warren's mapping staff. Roebling had enlisted in 1861 and rose from private to lieutenant colonel over the course of the war. When he met General Warren in June 1863, the younger Roebling was under orders to procure the best maps of the Pennsylvania-Maryland border region. Roebling accompanied Warren to Baltimore (where the general married Emily F. Chase).[64] On Little Round Top during the second day of Gettysburg, Roebling and Warren became heroes holding the Union line. Roebling eventually became the aide-de-camp to Warren. Emily Warren met Roebling at the Second Corps officers' ball in February 1864. They began their lifelong relationship there in Washington, DC, courting at the Chase residence in Baltimore, and marrying in Cold Spring on 18 January 1865.[65] So, by the end of the war, both the general and his sister had found suitable spouses.

Emily Warren Roebling learned much about engineering and suspension bridge construction from her husband as she wrote the orders he issued to the engineers building the East River Bridge. She demonstrated a keen sense of mathematics and quickly comprehended the engineering principles behind those written orders.[66] She maintained detailed scrapbooks as a personal record of the great engineering accomplishment for which her husband was the "captain."[67] One gets a sense that if Emily had been born a century later, and come of age in the era of equal rights for women, she would have been the person in charge of the bridge, making key decisions. She was driven to succeed at anything she pursued in her life. For example, in 1899, at the age of fifty-six, she earned a law degree from New York University. At the graduation ceremony in Madison Square Garden, she delivered one of the valedictory addresses.[68] But as a woman of her time, her creative outlet beyond the household was the hundreds of letters she drafted for her ailing husband. Her husband was her teacher, and she proved to be an exemplary pupil, putting his directions into clearly written instructions over most of the last decade of bridge construction.[69]

Emily also managed her social relationships with the greatest sense of propriety and a fair amount of dedication and loyalty. One of the more capable engineers working for the Roeblings was William H. Paine. Paine rose to the rank of colonel as a topographer for the Army of the Potomac during the Civil War.[70] Like Washington Roebling, Paine had served honorably on the staff of Emily's brother, Gen. G. K. Warren.[71] After the war, Paine moved to Brooklyn and joined the Roeblings' New York Bridge Company in 1869, where he made substantial surveys for the bridge's caissons and towers.[72] Emily Roebling came to depend on Paine above the other engineers, especially during those periods when her husband was most affected by the bends (caisson's disease). Noting that she looked upon Paine as "belonging" to her "family," Emily often requested Paine's presence when meeting with the other engineers.[73]

Emily and Catherine Jones Paine, Paine's second wife, also exchanged correspondence and gave Christmas presents to each other's children.[74] On her trademark flowered stationery, Emily ensured that the Roeblings' social obligations were met just like her husband's bridge instructions. When "Messer [*sic*]" and Madame Ferdinand de Lesseps paid a visit to the Roebling home in March 1880, Emily made certain that Catherine Paine was among the guests present to meet the famous French canal engineer and his wife.[75] Construction milestones became great celebrations for the engineers and their spouses alike. At an 1872 launch of the "New York Caisson" into the East River, Emily Roebling, Catherine Paine, several wives of the bridge officials, and "many others of note" made the trip to see the great spectacle.[76] The relationship between the Roebling and Paine families was a tapestry of shared war experience, engineering ambition, and community support, and the building of the East River Bridge was a validation of the life they pursued together in postbellum New York.

Again, what made this coterie of engineers and their families possible was those connections first fostered in war under the leadership of General Warren. As a wartime bride, Emily experienced anguish and worry about loved ones in harm's way not only for her brother, but also for her soon-to-be husband. Once in Brooklyn, she created a social world that expected manners and courtesies valued by a Victorian sense of decorum, a world reflecting the demure behavior she saw growing up in Cold Spring and at West Point. In this light, she was able to expand her knowledge and responsibilities, earning the respect and trust of her husband and those engineers serving under him. By the time the Brooklyn Bridge opened in 1883, New Yorkers generally accepted the impact of Emily's contributions to the entire project, even to the point where Brooklyn mayor Seth Low marveled at the power she wielded in the name of her infirm husband.[77]

While West Point and the Civil War proved to be important influences for connecting the most talented and capable engineers of New York, especially those working on the Brooklyn Bridge, West Point alumni did not always see eye to eye with their engineering brethren. In May 1879, Slocum leveled the most serious of accusations against Roebling and his engineers, including William H. Paine. He accused the bridge engineers of taking bribes from the Chrome Steel Company when Roebling changed the iron truss work into steel truss work. Roebling's actions created great suspicion of bribery and favoritism.[78] Since coming to Brooklyn, Slocum had had to be careful that he not appear unduly influenced by Boss McLaughlin's Brooklyn Ring. Also, with the construction of the Great Bridge repeatedly delayed, Slocum did not want to appear as part of the corruption or the problems associated with the project.[79] Although Roebling had served in the Civil War under fellow West Point graduates McDowell and Pope, and with G. K. Warren at Gettysburg, Slocum believed that his political status relied more on the public's perception of his character than on the ties of school and battle. What is more, Slocum took the word of Roebling's accuser as fact.[80] A special designated committee investigated the charges and cleared John A. Roebling's Sons Wire Company and its associated engineers of any wrongdoing. Slocum failed to apologize when the committee reported its findings, but in a later scandal, he did vote to keep

Washington Roebling as the chief engineer. Even so, the damage to Slocum's standing with the Roebling family was done, and they never forgave him for his part in making the initial accusations.[81]

Emily Roebling's legacy and place in New York's history remained secure. Her husband, who was so ill during construction of the bridge, outlived Emily by over two decades (he died in 1926). He and their children ensured that Emily's contributions would not be forgotten.[82] In 1953, the Brooklyn Engineers Club placed a bronze plaque dedicated to Emily on the east tower of the bridge. The plaque memorializes her with the words, BACK OF EVERY GREAT WORK WE CAN FIND THE SELF-SACRIFICING DEVOTION OF A WOMAN.[83] During the centennial celebration of the Brooklyn Bridge, Emily was a central figure, as well as in Ken Burns's 1982 documentary film about the bridge.[84]

In contrast to the attention given to Emily, the men of the East River Bridge are memorialized collectively on two identical plaques, one on each tower of the bridge. Among the sixty-eight names is Henry W. Slocum, listed with all of the trustees who were on the board at the opening in 1883.[85] Given the course of Slocum's political machinations during the bridge's construction, he is appropriately remembered as one of many who contributed to the completion of the Great Bridge.

SLOCUM AND THE PARTY MACHINE

Perhaps the main impediment to a greater legacy for Slocum was his inability to stand out on his own, both in war and in peace. Slocum's accomplishments generally followed on the actions of those above him, with McClellan, Meade, and Sherman being the most convincing cases. Observers viewed Slocum, as a Brooklyn politician, to be a pawn in William Kingsley's political agenda. When the board of trustees for the East River Bridge threatened to remove the incapacitated Roebling as chief engineer in September 1882, Slocum voted with Kingsley during the crucial vote to keep Roebling on.[86] Slocum's vote came as a surprise because he had been so vocal about Roebling's poor health and lack of capacity to lead construction.[87]

As had occurred in 1879, the 1882 crisis saw Slocum side with the Brooklyn politicians over Roebling the engineer. Brooklyn mayor Seth Low had set his sights on the New York gubernatorial race and needed the bridge completed during his tenure. That summer, Washington and Emily Roebling vacationed in Newport, Rhode Island, where Emily's brother was living. Slocum demanded that Roebling appear before the bridge trustees to explain the latest delays in construction. When the chief engineer declined in order to continue his vacation, Mayor Low made a secret journey to Newport and insisted in person that Roebling step down as chief engineer. Emily steadied her husband after the mayor stormed out of the vacation house, and Roebling stayed on as chief engineer through the completion of the bridge. Low's heavy-handedness might have influenced Slocum's decision to keep Roebling, but in the eyes of the Roeblings and their supporters, Slocum was a politician as malleable and corrupt as the rest.[88]

Also contributing to Slocum's mixed legacy was his tendency to speak his mind with little or no regard for the positions of his party or those around him. One would think that General Slocum would have learned from his experience as military governor of the Vicksburg district in 1865, but he continued to speak his mind in political matters as a lawyer and Brooklyn politician. He crossed Boss McLaughlin by openly criticizing the Brooklyn Ring's practice of handpicking Democratic candidates to run for office and for appointed positions. Slocum believed that expertise should trump party loyalty.[89] In 1875, he was quick to join in the reform of city politics following the removal of Boss Tweed from Tammany. Slocum attempted to do the same to McLaughlin in Brooklyn, but ultimately Boss McLaughlin was too strong to be overthrown.[90] McLaughlin's support ran deep among the immigrant and working-class communities of Brooklyn.[91] When Slocum was appointed commissioner of Brooklyn's public works in 1876, he made substantial efforts to reform the department. Within weeks of taking the position he published a letter in the *Brooklyn Daily Eagle* indicting the whole department for corruption and testified about a pervasive spoils system during the subsequent investigation.[92] The following year, Slocum cut the number of department employees and reduced employee salaries, including his own of $7,000 by $2,000.[93] He resigned two years later after repeated incidents of patronage and cronyism by department supervisors.[94] In the final evaluation of Slocum as a Brooklyn politician, one could say that he pursued his goals while sensitive to the public's perception of his motives, but he lacked the support and wherewithal to rise above the power and influence of the political machine.[95]

Until the end of his life, General Slocum supported causes that he believed were just. In the 1880s, he was one of Fitz John Porter's principal advocates during Porter's controversial court-martial for Second Manassas.[96] Slocum was a consummate favorite among his fellow Civil War veterans, who honored him with a monument on Culp's Hill at Gettysburg.[97] In spite of Slocum's popularity, his fellow veterans were also subject to Slocum's sense of fairness, even when it ran counter to their interests. In 1890, General Slocum used a veterans' reunion as a forum to sound a "warning against extravagant pension legislation." In his mind, the government had done enough for Union veterans, and to extend or increase payments to them would make them "mendicants of [themselves]."[98]

Whether Slocum was adhering to a code of honor or pursuing a political course that was naïve remains debatable, especially when considered within the limited scope of his life. However, when Slocum's judgment and decisions are viewed in the context of the political scene in postbellum New York, they signify a general limit to what the former West Pointers achieved in Gotham. Similar to Viele in New York, Slocum was able to influence the political discourse and affect the city's development, but only as a secondary player. Military expertise, wartime notoriety, and a network of West Point and army peers collectively enabled Slocum to enter Brooklyn's and New York's competitive postwar political environment. In the end, though, the Brooklyn machine curtailed any real power Slocum could have hoped to attain as a public servant in Brooklyn.

MCCLELLAN'S INTERLUDE AT THE DOCKS, 1870–1873

Across the East River, Boss Tweed's Tammany machine proved to be too much for another West Pointer and Civil War hero, Gen. George B. McClellan. Unlike General Slocum, though, McClellan was a formidable military commander and political power during the Civil War. McClellan was no stranger to New York. In the 1850s, George Templeton Strong often visited West Point to escape the city in the summer, staying at Cozzen's Hotel and visiting with Jacob W. Bailey and his family.[99] Strong traveled both by rail and by steamboat to enjoy the "social life" of summer at the Military Academy. He also got to know the professors well, including Mahan, William Bartlett, and, especially, Bailey. When he was a young captain in the Corps of Engineers, McClellan's exploits fighting Native Americans in Texas made him famous among West Point's faculty. Strong remarked in his diary that "Captain McClellan . . . was a very fine fellow whom we all liked extremely."[100] Strong's opinion of the young officer was a common one before the Civil War, and one that helped keep Mc-Clellan popular in New York through the war. McClellan's time as commissioner of the docks for New York City proved to be an intriguing interlude for such a luminous figure of the nineteenth century.

Graduating second in the USMA class of 1846, "Little Mac," as McClellan was sometimes called, served in the Mexican-American War as a lieutenant in the Corps of Engineers.[101] Subsequently, his duties included surveying prospective routes through the Cascade Mountains for the Transcontinental Railroad and reconnoitering Santo Domingo as the United States considered annexing the Dominican Republic.[102] In 1856, Secretary of War Jefferson Davis appointed McClellan to the Delafield Commission, a U.S. Army delegation sent to observe the Crimean War and bring back lessons learned.[103] By 1857, McClellan decided to try his hand at business, resigned from the army, and put his surveying and railroad expertise to use for the Illinois Central Railroad. From 1859 to 1861, McClellan earned a lucrative $10,000 a year as president of the Ohio and Mississippi Railroad. Living in Cincinnati, Ohio, he supervised the railroad between Ohio and St. Louis, protecting the investments of New York businessmen William H. Aspinwall and Samuel Barlow.[104] All the while, he continued to study and read about military tactics and strategy. On the eve of the Civil War, "Little Mac" had become one of the nation's most renowned military thinkers and successful railroad executives.[105]

McClellan revealed his political inclinations as a conservative Democrat when he openly supported Stephen Douglas in the 1858 senatorial election. He provided Douglas a private rail car to use for campaigning around Illinois. Apparently McClellan was not impressed by Lincoln's legal representation of the Illinois Central Railroad in the late 1850s. Moreover, "Little Mac" thought Lincoln lacked "strong character" and the resolve to make difficult decisions, an opinion that lasted throughout the Civil War.[106] In spite of McClellan's prewar opinion on the character and personality of Lincoln, he agreed to command the Army of the Potomac when the commander in chief called upon him in July 1861.

McClellan's record as commander of the Army of the Potomac certainly failed to meet the leadership's expectations of an officer whose nickname was "Young Napoleon." For over fifteen months, Lincoln urged McClellan to crush the Confederacy by defeating Lee's Army of Northern Virginia and capturing Richmond. Under McClellan, the Army of the Potomac failed to accomplish either. The Peninsula Campaign stalled four miles outside of Richmond, and

Lee's army escaped after the Battle of Antietam in September 1862. By November, the president had had enough of the general's overly cautious approach to defeating the rebellion and relieved "Little Mac" of his command. McClellan withdrew to his home in Trenton, New Jersey, to consider his future.[107]

While there were calls for McClellan to retake the field of battle during the remainder of the Civil War, he stayed in New Jersey preparing for a campaign as the 1864 Democratic presidential nominee. In New York, the former Union Army commander had a wellspring of Democratic support and became the party's favorite candidate, including both its Peace and War factions.[108] Chief among Gotham's Democrats backing McClellan were August Belmont and Samuel Barlow. Not only did the young general have military star power, but his opposition to the "radical" policies of Lincoln and the Republicans was well-known.[109] During the Draft Riots of 1863, McClellan and Barlow counseled Gov. Horatio Seymour on how to quell the crisis from the governor's command post at the Saint Nicholas Hotel.[110] Throughout 1863, the Peace Democrats, led by Fernando Wood, became more adamant in their calls to end the war through a settlement that would circumvent emancipation and allow the Confederate states to reenter the Union on more lenient terms, or instead, to remain a separate nation. However, Belmont and Barlow were War Democrats who wanted to see the Union preserved, but under much less stringent terms than those demanded by Lincoln and his administration. The challenge of the McClellan candidacy, therefore, was to gain the support of both factions of the Democratic Party in time for the national election.[111]

In this setting, McClellan attempted to manage his campaign messages to meet the pro-Union planks in the Democratic platform. At a speech in West Point in June 1864, the only one of his campaign, his message was one of union and support for the Constitution, saying that "conciliation, common interest and mutual charity, had been the foundation and must be the support of our Government."[112] His remarks were equivocal enough to secure him the nomination at the Democratic National Convention in Chicago, but McClellan turned out to be as inept in presidential politics as he had been while leading the Army of the Potomac. Going into the November elections, the Democrats split their support between McClellan and Clement L. Vallandigham's peace platform, a position that McClellan did not endorse, even after winning the nomination.[113] Some prowar Democrats further complicated the Democratic position by supporting Lincoln under the banner of the National Union Party.[114] No one campaigned harder for McClellan and the Democrats than

August Belmont, but even the New York banker could not counter the calamitous outcome of the Chicago convention. The general and the Democrats lost in a landslide.[115]

After McClellan lost his bid for the presidency, he returned to the business world, seeking to become president of the Morris and Essex Railroad in New Jersey. He had hoped that the industrialist Abram Hewitt, who later became mayor of New York, would negotiate with the board of directors to secure him the job, but the general's reputation had suffered greatly. According to Hewitt, the board of directors was not comfortable with McClellan as their company head for fear that it would jeopardize company relations with the government. McClellan complained to Barlow that he was being punished "merely because a great & honest party chose to make me their leader." McClellan then set off on 25 January 1865 to tour Europe, financed by his rental income and railroad stock invested in a mining venture.

McClellan traveled through Britain and the Continent. The highlight of his time abroad was meeting the great general Helmuth von Moltke, chief of staff of the Prussian Army. On 29 September 1868, McClellan sailed back to the United States at the height of the 1868 campaign. In New York and in Philadelphia, McClellan received wide and enthusiastic popular support. He even considered running with Democratic nominee Horatio Seymour against U. S. Grant and the Republicans. Still smarting from the defeat of 1864, McClellan chose not to pursue another nomination.

While in Europe, McClellan worked for Edwin A. Stevens, looking for foreign buyers for the Stevens Battery, a design for a doomed oceangoing ironclad over four hundred feet in length. Stevens died in August 1868, stipulating in his will that $1 million would go to complete the battery if McClellan remained its chief engineer. McClellan moved to Hoboken, New Jersey, to fulfill Stevens's wish, but the project ultimately turned out to be a failure, going to the scrap heap in 1870.[116] After the demise of the Stevens Battery, "Little Mac" had to be cautious with whatever he chose to do next. Still relatively young at forty-three, he needed a new position to sustain the comfortable lifestyle and status to which he and his family had grown accustomed.

McClellan received several offers to lead companies or institutions. Most notably, the University of California's board of regents offered him $6,000 a year to become president of the university.[117] In their campaign to lure McClellan westward, they promised to increase his salary to meet his current income and appealed to his pride by hinting at the status to be gained through the position.[118] Settled comfortably in New Jersey, McClellan declined this and other lucrative offers that would require him to relocate from the New York area. Not having any better prospects, he decided to become chief engineer of New York City's Department of Docks in July 1870.[119]

In other treatments of General McClellan's life, his tenure as chief engineer of the Docks Department draws limited attention in comparison to his military service or political campaigns. However, within the context of postbellum Gotham, McClellan's time at the Docks Department reveals the confluence of several dynamics present in New York after the war, mainly through his combination of engineering expertise and his political capacity in using this position.[120] McClellan's expertise in leading large projects such as railroad construction or territorial surveys made him a logical choice for Mayor Oakey Hall and the city in 1870. Furthermore, McClellan's three-year tenure as chief engineer was another attempt by engineers to create order in a city bursting with uneven growth and development.

As the Port of New York rapidly expanded after the war, merchants, waterfront landowners, and commercial shipping owners sought a single state-run harbor commission that could regulate Manhattan's and Brooklyn's waterfronts. Competing merchants and marine businessmen devised a privately owned New York Pier and Warehouse Company to construct a standardized port system "of stone quays, iron piers, and dockside stores with steam hoists and railroads." Not wanting to lose control over such a lucrative source of patronage, Tammany changed the scope of the plans and pursued a larger, more encompassing public dockside project for Manhattan's entire waterfront.[121]

Boss Tweed backed creation of the Department of Docks, and through Mayor Hall he charged McClellan with designing a master plan to be backed by public funds. In addition to hiring McClellan, the board sent engineers to London and Liverpool to find the best dock and pier system to adopt for New York.[122] By November 1870, General McClellan had proposed a series of uniform piers that would ring the shoreline of Manhattan from 61st Street on the West Side "to Corlears Hook" in the East River. The piers would connect to a "new bulkhead . . . around the city" that would form a new street to be called "River-street." Also on the new road was to be a railroad, wharf warehouses, and steam-powered cranes (which imitated the plan devised by New York Pier and Warehouse Company).[123] Priced at $1 million per mile of wharf line developed, the waterfront development was to be a multi-million-dollar boon for the city. The dock commissioners proposed to fund the massive masonry and iron project with viaduct railroad funds previously approved. They thought that a railroad ringing the city would be more beneficial than one built above it.[124] McClellan's plan in effect was another considerable effort to reform the forces of increasing disorder that he and others saw after the Civil War. By 1872, the massive project stalled, overcome by the diverse interests of all involved—shipowners, merchants, landowners, insurance firms, ferry companies, corporations, and canal owners—and the city waterfront remained a mix of squalor and uneven development.[125] As chief engineer of the docks, McClellan failed to attend any of the hearings where waterfront proponents proposed their visions for the new dock system. He more than likely took a practical approach and realized how unfeasible the proposals were. Ultimately, he lacked the power to subdue the myriad of players whose livelihoods depended upon New York's shipping and port activities, suggesting the limits to his expertise and influence.[126]

During this same period, Tammany Hall's grip over city affairs loosened as the corruption of the Tweed Ring came to light. When Tammany's Richard Connolly resigned as New York's city comptroller in September 1871, Mayor Hall requested that McClellan replace Connolly.[127] McClellan believed that he was honor-bound to take the post and help the city recover from the financial crises created by Tweed and his cronies. Longtime counselors and friends such as William Aspinwall, William Hunt, Gen. William Wallace Burns, and Gen. William Averell all strongly advised him not to take the comptroller position. In the autumn of 1871, the extent of Tammany corruption and "dirty politicians" was just becoming clear.[128] For the famous general to take such a professional risk would have more than likely ended any further political aspirations he had at that point. McClellan followed the advice he was given and stayed on at the Department of Docks a little longer.

By April 1873, McClellan left public service again for the private sector, resigning to start his own engineering company, Geo. B. McClellan & Co., Consulting Engineers and Accountants. Although he submitted a letter of resignation prior to starting this new venture, McClellan was already serving as president of another private company, the Atlantic and Great Western Railway Company, where he was earning upwards of $15,000 a year.[129] At his new firm, he focused on securing or insuring the investments of European clients in American railroads in an age of seemingly unbounded capitalism. Living in New Jersey, he remained active in the state's Democratic Party. In 1877, Mc-

Clellan was nominated as the Democratic candidate for governor and won. From 1878 to 1881, he reduced New Jersey's deficit by 23 percent and ended direct state taxes on the people.[130] It was the highest public office McClellan would attain.[131]

Like Henry Slocum in Brooklyn, McClellan's experiences in New York and New Jersey were mainly political in nature. Both men came to the city to revitalize their post–Civil War careers and, in the process, to diminish or suppress blemishes on their war records. Granted, McClellan's failures on the battlefield had greater strategic and national implications than any error of Slocum's. Regardless of the magnitude of either officer's shortcomings in war, however, their fame and military reputations were prizes for the machine politicians in postbellum New York and Brooklyn. With antebellum social and political ties to New York, these two West Point generals relied upon their fame and reputations as well to leverage their way into Gotham's dynamic political and social scene. When compared to Slocum, McClellan was superior at promoting himself, an art he learned early on as a young officer and mastered during the Civil War.

What is also noteworthy about McClellan's experience was his ability to market his engineering expertise before and after the war. Unlike Slocum, who forged his civilian career only around the law and not engineering, McClellan exploited his engineering experiences for distinction in and out of uniform. During his tenure as chief engineer of the Department of Docks, he tried to apply his grand vision for the city's port enterprise and infrastructure, but only so long as it served his needs. The absence of a newly constructed system of stone bulkheads lining Manhattan's shores in 1873 ensured that McClellan's engineering legacy in New York would remain less than memorable. McClellan had no structure or engineering creation to attach to his legacy. In short, he had no Central Park, Croton Aqueduct, or Brooklyn Bridge.

SETTING THE STAGE FOR GREATER NEW YORK CITY—
THE METROPOLIS LINKED

When the East River Bridge opened on 24 May 1883, it was a celebration of the symbolic and tangible connection of two great cities, New York and Brooklyn. It was also a truly modern marvel, the "Eighth Wonder of the World." The

Brooklyn Daily Eagle proclaimed it "the Greatest Engineering Feat of the Century."[132] Front and center for all to revere stood a monument to the prowess of New York's engineers. Politicians may have been the focus of the crowd's attention as they led the public celebrations, but the engineers were the real heroes of the day. Governor Grover Cleveland, Mayor Seth Low, James Stranahan, William Kingsley, and Henry Slocum each made their way to the Roebling house to congratulate Washington Roebling personally for the bridge.[133] Its opening, more than any other event, marked the first real step toward the consolidation of the five boroughs into Greater New York.[134]

Political consolidation was some fifteen years away, but engineering projects continued to connect the metropolis. Before the end of the century, New York built three new bridges connecting Manhattan to Harlem and two more over the Gowanus Canal, and commissioned the Williamsburg Bridge as a second East River crossing.[135] While municipal leaders figured out how to reach an agreement that all boroughs could live with, engineers and builders met the increasing demands for commercial infrastructure in Gotham.

In addition to bridges, there were other internal improvements to be completed around the port, including the removal of the Hell Gate rocks in the East River. Since the first European settlers voyaged up the east side of Manhattan Island, Hell Gate had been a treacherous obstacle to navigation. Generally located where the present-day Robert F. Kennedy–Triborough Bridge spans the river, Hell Gate was also where the East River, the Harlem River, and the Long Island Sound merged, creating treacherous currents and conditions, especially during changing tides. The currents could not be manipulated, but if the rock obstacles were to be removed, vessels would have more space to maneuver and negotiate the most perilous portion of New York's waterways. By the mid-nineteenth century, Hell Gate had claimed hundreds of ships, with thousands more running aground.[136] In 1852, Congress first appropriated funds ($20,000) for the Army Corps of Engineers to remove the rock.[137]

Starting that year, the Corps began blowing up rock obstacles in the waterways, beginning with Hell Gate.[138] For nearly seven decades, the army blew up rock and excavated debris from Hell Gate. In the process of clearing the hazards at Hell Gate, the Corps of Engineers also experimented with "torpedoes," or explosive water mines that could be used against enemy ships. Maj. Henry L. Abbott, USMA class of 1854, led the development of antisubmarine mines at Willets Point, using the tidal currents and changes in Hell Gate as a testing ground for the mining system.[139] New Yorkers grew accustomed to the sound of dynamite explosions in the East River. The most famous explosion was that of 10 October 1885, when Gen. John Newton successfully blew up the main obstacle—"Flood Rock," as the nine-acre outcropping at Hell Gate was known. General Newton's "scientific violence" was felt all over New York and as far away as Princeton, New Jersey.[140] For Newton, this spectacular explosion was the culmination of eighteen years of improving New York's rivers and harbor.[141]

John Newton was the quintessential West Point army engineer. He was the top cadet in the USMA class of 1842 for all four years, and was commissioned accordingly as a lieutenant in the engineers upon graduation. Newton served in key engineering assignments throughout his career. From surveying rivers and harbors to building the nation's coastal fortifications to teaching engineering at West Point, Newton lived up to the potential he demonstrated as a cadet. During the Civil War, he participated in major battles and even commanded a Union division at the Battle of Gettysburg. By war's end, Newton was one of the most capable army engineers, and the Corps of Engineers therefore made him responsible for fortifications and improvements in New York. Eventually, Newton would retire from the army in 1886 as chief of engineers.[142]

For New York, Newton's greatest contribution lay not so much in removing the Hell Gate obstacle, but in carrying out explosions and clearing rock debris. In the 1850s and 1860s, the initial blasting efforts came from lowering a canister of gunpowder down onto the rocks under the water and exploding it there. This method was unpredictable and expensive and usually shattered windows on buildings adjacent to the water.[143] In 1867, Alfred Nobel's invention of dynamite liberated the Hell Gate "rock blasters" from having to use the more unpredictable black powder explosions. When Newton took over the clearing of Hell Gate in 1867, he devised a caisson and drilling machine system that allowed contractors to drill into the rock underwater and emplace the dynamite charges into the rock.[144] Later he improved the machine with "a steam-driven scow" that could drill down nine feet quickly and repeatedly. Newton's method became so efficient that he charged 50,000 pounds of dynamite on Hallet's Point Reef.[145] With the Hell Gate operations, Newton was so confident in the outcomes of his controlled explosions that he allowed his two-year-old daughter to plunge the charging switch during the operation. This helped to assuage New Yorkers' concerns over the whole process. Furthermore, Newton's expertise became more and more evident over time with each successful blast.

Thus, by the end of the century, the East River was easily navigable through to the Harlem River and Long Island Sound at a relatively small inconvenience to New Yorkers.

With an ever improving harbor and waterway system, New York and the four other boroughs became more integrated into a single major seaport. As the Army Corps of Engineers led the effort on the rivers, and with the coastal system of lighthouses and channel surveys established nationally, New York benefited directly. The completion of the Harlem River Canal was the last missing piece of New York's waterway network. This canal enabled ship traffic to circumnavigate Manhattan by connecting the Hudson and East Rivers at the island's north end. On 17 June 1895, with the fanfare and ceremonial festivities that New Yorkers had grown accustomed to at the end of the century, Egbert Viele led the celebration opening the canal.[146] Abounding with two parades, a flotilla of vessels led by a navy cruiser, and a ceremonial "wedding of the waters," the occasion was about more than the opening of a canal. It was another ritual sanctifying the engineering feats that New Yorkers believed were leading them and the nation to a greater destiny. No longer did Gotham's greatness just come from public parks, better sanitary conditions, and unbounded commercial activity. At the end of the century, magnificent bridges and increased commercial capacity were the measure of the metropolis's greatness.

CONCLUSION

In many ways, engineers were building the city Andrew Haswell Green had dreamed of for New York. During the aftermath of Tammany's (and Tweed's) fall in 1871, Green became the city comptroller, leading a crusade of civil reform while promoting budget austerity and cutting social services programs. Although the New York State Legislature created a new city charter in 1873, the form and function of municipal departments remained much the same. Green came to the conclusion that there were efficiencies to be gained by consolidating various agencies and departments of the boroughs into a single municipal authority.[147] According to Green's thinking, the city commissioners and department heads could function under one government with less potential for corruption and graft. Party machine interests in Manhattan's Lower East Side would not be the same as the interests of Laughlin's working-class supporters in Brooklyn. A stronger, centralized municipal power over a consolidated Greater New York could reform bad government and resolve social ills. However, resistance to consolidation grew in the boroughs as the pressure of economic stress, exponential population growth, and the precariousness of Gilded Age business forced New Yorkers to focus locally on protecting their own interests.[148] Engineers were certainly limited in what they could do with regard to achieving reforms and realizing ideas of social justice, but the bridges, river channel maintenance, and port improvements they made increased the interests the boroughs held in common.

In other ways, engineers were building a New York that differed from Green's vision. Improved infrastructure meant more commercial activity and, in postbellum Gotham, capitalism reigned supreme. New York's demand for labor and business space quickly outpaced available land on Manhattan in the second half of the nineteenth century.[149] Here, too, the engineer community literally rose to the challenge by building skyscrapers that increased business space exponentially. Railroad architect Bradford Gilbert built an eleven-story, 158-foot-tall building on Broadway in 1888, starting the era of tall buildings.[150] The Brooklyn Bridge and the elevated railways brought millions to and from work each day. Whether Green could get the boroughs to consolidate or not, New York's commercial growth was going to continue on a pace of its own into the next century. Ultimately, New Yorkers voted for the consolidation referendum in November 1894 that led to the 1898 merger, because the potential benefits of belonging to a single metropolitan New York outweighed the risks of remaining separate from Greater New York.[151]

By coming to New York after the Civil War, West Pointers pursued the "manifold possibilities" and lived the "intense" life that Theodore Roosevelt richly described in his 1891 history of the city.[152] While it is true that personal interests frequently were a motivation to settle in New York or Brooklyn, the West Pointers highlighted in this book each sought to contribute and be a part of a greater enterprise. In Emily Warren Roebling's case, because she was the sister of a graduate, the city benefited from her education and association with West Point. For Slocum and Roebling it was the Great Bridge, for McClellan the waterfront development of New York, and for Newton the city's waterway system. Although Newton was still in active service while removing Hell Gate, his posting in New York was a plum assignment for an army engineer, especially after the Civil War. Whether or not it was a conscious goal, the professional pursuits of West Pointers and Emily Roebling were manifested in the infrastructure and conditions for Gotham's consolidation in the 1890s.

Twenty-seven years after George Templeton Strong compared the parks of Brooklyn and New York using the pronoun "we" for Central Park and "it" for Prospect Park, both parks could be claimed for all New Yorkers.[153] Gotham in 1898 was truly a different city than it had been in the first decade after the Civil War, but so too were the graduates of West Point living there at the end of the century. Instead of building parks, bridges, and shipping channels, the "old grads" pursued the creation of memorials and monuments to remember the great sacrifice of the Civil War and the triumph of the Union. Grant's Tomb, as well as statues to Sherman, G. K. Warren, and even Slocum were representative of the moment.[154] These monuments commemorated the most pivotal experiences of their lifetimes, but they also marked the end of an era for West Point in New York. As the generation of Slocum, Warren, and Newton faded into memory, so too did the influence of West Point graduates on the national scene.

The population of the United States, as well as that of New York, tripled in the second half of the century. The nation was developing—industrially, geographically, demographically—at such a rate that the alumni of the U.S. Military Academy were too small a body of experts to meet the demands of the country as well as metropolitan New York.[155] Whether or not it was apparent at the time, the Civil War generation of West Pointers had had their moment. The network and connections they had made as cadets, as army officers, and as figures in postbellum Gotham would not be as effective for the next generation in the new metropolis. Instead, power in Greater New York emanated from two bases: Richard Croker's Tammany Hall, and the reformers led by Seth Low who tried to curtail the chieftain's reach.[156] While there were West Pointers who had graduated after the Civil War and attained some degree of success in New York, none was as effective at facilitating change as the alumni who had served in the war.[157]

Forty years after the Civil War, the closest connection to West Point and leadership in the city was the mayoralty of George B. McClellan Jr., the son of the general. From 1904 to 1909, he led City Hall, initially with the support of Tammany boss Charles Francis Murphy. Mayor McClellan proved to be too honest for Murphy by insisting on working with the state legislature in the expansion of the subway and improvement of the water supply through a public works project in the Catskill Mountains.[158] The general's son left a lasting legacy in New York with these public works projects, but his accomplishments were due more to his own efforts and abilities than his father's renown.

Still, as Greater New York became increasingly subject to the campaigns of Progressive reformers like Theodore Roosevelt, Jacob Riis, and Seth Low, there were antecedents in the pre-consolidation era highlighted by Military Academy graduates who had led the municipal departments. In their capacity as commissioners of the Police Department and Public Works, they attempted to control the scope of change in the city between 1865 and 1898. Those efforts varied in outcome and effect, but they were just as important to the network of civil engineers, West Pointers, and municipal leaders that forged the political, economic, and social bonds in the city of five boroughs.

7

REDEMPTION IN
POSTBELLUM GOTHAM

In short, the most important lesson taught by the history of New York City is the
lesson of Americanism,—the lesson that he among us who wishes to win honour
in our life, and to play his part honestly and manfully, must be indeed an American in spirit and purpose, in heart and thought and deed.

—THEODORE ROOSEVELT, New York, 1891

THE ORDEAL OF FITZ JOHN PORTER

"Cleveland had signed the bill" was all it said. Tears welled up in Fitz John Porter's eyes as he read the note at his desk in the office at 300 Mulberry Street.
A messenger had delivered the news that had eluded him for nearly a quarter
century. Porter, a commissioner of the New York Police Department, was once
again a colonel of infantry in the U.S. Army, overturning his court-martial
conviction of 1863.[1] There had been moments over the years when he thought
his conviction would be overturned, but each of those attempts ended with dis-
appointment. This time, in June and July of 1886, Congress debated and finally
passed the act that called for his reappointment and full pardon. The fact that
President Grover Cleveland waited almost a week to sign the bill heightened
Porter's anxiety. His reinstatement was a vindication after years of personal
sacrifice, financial hardship, and political lobbying to clear his name. Until that
summer's day in 1886, postbellum New York was his personal purgatory on
earth. Even though he had been widely exonerated by a military commission of
inquiry, his Democratic peers, and much of public opinion in the city, without

the official pardon he could never adequately restore his reputation among his generation.

In January 1863, as General Porter, he had been cashiered through a court-martial following his actions at the Battle of Second Manassas in August 1862. The Union commander, Gen. John Pope, blamed Porter for the Union defeat, charging that he did not follow orders and left the Union flank vulnerable to attack. History and the Congress eventually demonstrated that General Pope's commands required Porter's soldiers, who were outmanned 3 to 1 by Stonewall Jackson's Rebel soldiers, to fight simultaneously in opposite directions— an unrealistic expectation in any battle. Porter was a victim of bad luck and poor leadership.[2]

Apart from the shame and degradation in status the verdict created for him and his family, General Porter could not serve in any federal office, including Congress, as a former officer with a court-martial conviction.[3] In New York, though, the dishonored former Union general could serve as an appointee of the city's elected officials. By any measurement, Porter, an 1845 graduate of West Point and a distinguished veteran of the Mexican-American War, had served admirably prior to the controversy that ensued after Second Manassas and, more importantly, his service in the Civil War was honorable in the eyes of New York Democrats. They saw him as a practical, if not a natural, candidate to appoint as a commissioner, one who would remedy those municipal departments tarnished repeatedly by Tammany scandal. Porter's controversial war record may have excluded him from serving in a federal capacity but, in 1884, it was not a concern for Mayor Franklin Edson. In fact, the Tilden Commission and the "reformed" New York Democrats resisted federal interventionism in the city, something that Tammany and anti-Tammany men could agree on. Electing and organizing the city's government was a local matter.[4] The *Evening Post* agreed and affirmed, "[The] new Police Commissioner is a gentleman, by education, the habit of controlling large bodies of men, and other valuable personal qualifications, is well fitted for the responsible position he assumes. The public may be congratulated on obtaining the services of so able an officer."[5] At least for Edwin L. Godkin and his fellow *Post* editors, the West Point graduate and former general had the education, expertise, and experience to reform the maligned Police Department and restore honor and dignity to "New York's finest."[6] Porter may have been unacceptable for national service, but he was Mayor Edson's ideal candidate to lead Gotham's police force.[7]

At first glance, Fitz John Porter's ordeal appears to be just a story about an unlucky Civil War veteran finding redemption in postbellum New York. However, when seen in the wider context of the city's postwar political environment, Porter's tale illuminates the city's increased reliance on veterans with leadership and engineering experience, many of them West Pointers. Municipal leaders seeking to maintain power and control over the growth of Gotham in the second half of the nineteenth century needed individuals with experience, expertise, and the right political affiliation. Porter fit the bill for all three qualifications as a Military Academy graduate, former Union general, and member of the Democratic Party. Unlike West Pointers who capitalized specifically on their engineering expertise in New York, the alumni who became commissioners in city government used their leadership experiences in the army to organize and lead New York's municipal departments and boards. The result was a dynamic relationship between the West Pointers and New York in which each benefited. City Hall found qualified, if not reliable, leaders for its departments, and the graduates achieved some level of status and, in many cases, redemption for past shortcomings.

WEST POINTERS AS COMMISSIONERS:
GENERALS IN GOTHAM

The politicians of Victorian New York were quite astute at gaining and preserving power through the party machines. However, whether Republican or Democrat, they were seriously challenged in their ability to govern and preserve order before, during, and after the war. Personal interests and greed usually won out over the protection of private rights and public interests.[8] Reconstruction-era local governments, both state and municipal, often became profit machines for the opportunistic. Corruption could be rationalized as a necessary exchange of favors to ensure profit for all on the take.[9] In the calls for municipal reform, leaders from both parties came to rely on the expertise of individuals who could plan, organize, and supervise the fundamental services necessary to keep Gotham running. As seen earlier with antebellum engineers in building the Croton water system and with Viele's topographical sanitation maps, West Point–educated engineers were effective experts at achieving order, or at least a perception of order, and at promoting the internal improvements desired by city leaders.[10] As the city's departments adapted and evolved between 1850 and 1890, so did the nature of the Military Academy graduates' involvement in meeting municipal responsibilities. For these men, the bonds created by shared cadet experiences and military service in the Civil War continued to play a role in the city that had emerged victorious in the wake of four years of national sacrifice.

The civil-military relationship between Lincoln and his generals during the Civil War informed the postbellum experiences of West Point veterans. They had all witnessed the parade of generals the president had gone through in looking for a military leader to achieve his political goals. If New York's mayors and political leaders were the civilian leaders, setting policy objectives for Gotham's rise, then its commissioners could be viewed as the "generals" designing and directing strategies to achieve those objectives. In this relationship, loyalty to the political party mattered as much as one's ability to apply expertise in overcoming the challenges of a modernizing metropolis. Over time, expertise became more important than party affiliation.[11] After the Civil War, Academy alumni tended to fill municipal appointments based more upon their ability to organize and direct the departments and less upon their engineering expertise. Some would still be influenced by the impulses of Tammany, but more would be instated like Fitz John Porter, as an expression of reform and progressive change. Politicians could point to the reputation and wartime service of an Academy graduate to show their political base as well as their rivals that a remedy was in the works.[12]

How the West Point officers migrated to postwar New York depended on where and with whom they had served in the Civil War. Relationships forged over the course of fighting the war often led to professional networking in postbellum Gotham. The fact that Henry Slocum and Porter had served under McClellan and that all were Democrats was not a coincidence. Like-minded men were drawn to each other. Furthermore, because Democrats dominated all but four of the seventeen mayoral elections after the war, Democratic veterans had a good chance of landing a well-paid appointment with the municipal government in New York regardless of their standing with Tammany.[13] Even when Radical Republicans carried the national election of 1868, New York and Brooklyn remained obstinately in Democratic hands.[14] Those West Pointers with conservative tendencies could apply their expertise to reforming the ills of industrialization and urbanization. War-forged camaraderie and deep Democratic ties were the rule and not the exception for Academy appointees, especially in the 1870s and 1880s.

As David Hammack succinctly describes the period from 1872 to 1886, every mayor elected in New York City was a "prominent merchant" in the Democratic Party. Known as the Swallowtails, these merchants, bankers, lawyers, and elite businessmen filled the void in municipal politics that had been created by Tweed's fall in the early 1870s. During the era of the Swallowtails, West Point alumni who came to the city tended to support and pursue Swallowtail agendas over Tammany's interests. Mayor Abram Hewitt was the chief Swallowtail politician, but wealthy banker August Belmont proved just as influential.[15] The contest between the business elite and Tammany Hall during the period was not always clearly delineated. John Kelly, who had succeeded Tweed as boss, operated as an effective machine leader, and by 1880 he helped Swallowtail candidate William Grace win the mayoral election. Six years later, Richard Croker was the Tammany chief, and he secured the election of Hewitt, which signaled an emerging political partnership between labor and capitalists in the city. By the mayoral election of 1888, Hewitt and the Swallowtails were out, and Tammany, under Croker, once again controlled City Hall.[16] However, during most of this time, business and commercial interests dominated municipal motivations, both private and public.

The shift of Wall Street capital from the antebellum cotton kingdom to the railroad-driven westward expansion of the nation contributed to the allure of postbellum Gotham for veteran West Pointers. If an Academy alumnus was not able to secure a commissioner post in the city, then he could offer his service and expertise to the businessmen backing the construction of railroads or mining ventures in the West.[17] Having proved their value in the antebellum railroad industry, the West Point men could continue to apply their engineering skills to the railroad boom during and after the Civil War.[18] George McClellan was able to tend to his railroad business obligations while he was head of New York City's Docks Department.[19] And Fitz John Porter found lucrative employment working for the Central Railroad of New Jersey in between his stints as public works commissioner and police commissioner. McClellan, Porter, and others like George S. Greene alternated between working for private railroads and in the public sector.[20]

This pattern of railroad employment was not limited to West Point graduates in Gotham. By one count, forty-nine graduates served as chief engineers and twenty-two as presidents of railroads across the nation prior to 1900. The postwar railroad boom was a national phenomenon, as speculators and businessmen, including West Point alumni, profited from the enterprise of connecting the country from east to west and north to south. Among them were Ambrose Burnside (class of 1847) and Horace Porter and James H. Wilson (both class of 1860).[21] Yet the concentration of railroad wealth in New York was unique, and therefore it empowered business tycoons and those who worked for them.[22] Funding for any railroad in the Midwest and the West usually originated with investments made in New York. Chicago, Cincinnati, Buffalo, and Memphis also benefited from railroad industry growth, but none could surpass Gotham in the Gilded Age.[23] Though some West Pointers were among the successful railroad men of the time, more found civilian employment, albeit less lucrative than running a railroad, working for the City of New York.[24]

THE STREETS DEPARTMENT, PUBLIC WORKS, AND "TREASON"

Among New York's ten original municipal departments formed in 1849, the Streets Department was one of the first to attract both engineers and optimistic reformers.[25] By 1870, the Streets Department had merged with the Croton Aqueduct Department to become the Department of Public Works.[26] West Pointers led these agencies at several points before and after the 1870 reorganization. Their tenures illustrate the dynamic between Academy alumni and service in the city's departments.

Two members of the USMA class of 1842, Gustavus Woodson Smith and Mansfield Lovell, had a long-term personal and professional bond that began as cadets, strengthened during military service, and played out in Gotham. Smith and Lovell graduated eighth and ninth in their class respectively, served in the Mexican-American War and, in the late 1850s, resigned from the army to work in New York's Streets Department. Smith was the streets commissioner from 1858 to 1861; Lovell was superintendent of street improvements in 1858, and then was Smith's deputy from 1859 to 1861. Their parallel experiences mimic those of other graduates examined in this study, until 1861, when Smith and Lovell, independent of each other, left New York to serve in the Confederate Army. In the antebellum city, West Point alumni who had served in the Mexican-American War were a welcome source of expertise. Similar to George S. Greene and Egbert Viele, Smith and Lovell could claim that they had learned civil engineering at the Military Academy and applied it during their army service. What makes Smith's and Lovell's stories even more intriguing is that they both eventually returned to New York after the war. Smith worked as a life insurance expert, and Lovell was an engineer for Gen. John Newton engaged in removing Hell Gate (Newton had graduated second in the class of 1842).[27]

How did two "traitors" from the Academy leave New York to fight against the Union and then, after the war, return to live in the city? First, they were proslavery Democrats before the war. Lovell openly supported the compromises that failed to prevent the South's secession. Like McClellan and Viele, they were conservative military men who supported conciliatory Democratic policies that would have preserved slavery in the Union. Smith's and Lovell's antebellum sentiments were similar to those of New York's Peace Democrats, who argued that New York's economic fortunes were tied to the southern "planter aristocracy."[28] Second, they both shared a vehement belief in the inferiority of African Americans and strongly resented the abolitionists' control over the Republican Party agenda. When Lincoln won the 1860 election, Smith and Lovell believed that the abolition of slavery would become national policy under the Republican administration. These two West Point alumni realized that their convictions would not allow them to serve in the Union Army.[29]

For G. W. Smith, the question of party allegiance was not simply a matter of supporting the Democratic platforms proposed by Tammany or Mayor Fernando Wood. The fact that Mayor Daniel F. Tiemann, who served between Wood's two terms, had made Smith the streets commissioner in November 1858 complicated Smith's later relationship with Mayor Wood in 1861. Tiemann, who was not a Fernando Wood man, had hired Smith because of his West Point education and experience in the Corps of Engineers.[30] Tiemann, a Free-Soiler, had become mayor when Tammany Democrats and Republicans combined as the People's Party to counter Wood.[31] When Wood won back the mayor's office for a second term in 1860, he wanted to remove Smith and install his brother, Henry, in the office of streets commissioner.[32] However, the board of aldermen was not receptive to the new mayor's desires and kept Smith on as commissioner, a decision endorsed in the *New York Times:* "In Capt. Smith we have had an officer against whose official purity not even slander has dared to wag a tongue. An educated engineer, he has done all that could be done for the City within his term of office, elevating those in his Department who deserved elevation on the sole ground of merit, and never hesitating to punish delinquents or malfeasants, no matter how strong might be their political backers."[33] Smith also took action to strengthen his position as streets commissioner. He fired John H. Chambers from the office of deputy collector of assessments because Chambers had supported efforts to unseat Smith.[34]

In January 1861, Smith's loyalties remained strong to the city and the non-Tammany Democrats but, a few short months later, the national crisis caused Smith to reconsider his situation. Also in January 1861, Fernando Wood hoped that he could take advantage of the antiabolitionist sentiments rising in the city and suggested that New York City secede in the event the South left the Union. Wood's ploy was more a way to break Manhattan away from the state government in Albany than it was to support the slave states. Smith saw through Wood's gambit and did not support it.[35] Ultimately, for Smith, it was the actual outbreak of war that led him to see that the crisis of the Union was not going to be resolved without further widespread war. As the secession crisis played out at Fort Sumter in the early spring of 1861, Smith suffered a stroke that confined him to his home for two weeks. By the end of July, Smith was well enough to follow his doctor's orders to travel to Hot Springs, Arkansas, for rehabilitative treatment. During the trip, he learned in Kentucky that the U.S. government was detaining individuals who openly supported the South. Secretly, Smith sent his wife Lucretia to St. Mary's, Canada, while the New York press and City Hall speculated as to his whereabouts and his allegiances.[36] Smith's defection to the Confederacy came on 20 September 1861 with a letter of resignation sent to the board of aldermen.[37] By the end of the month, Jefferson Davis accepted Smith's offer to serve in the Confederate Army and made him a corps commander.[38]

For Mansfield Lovell, Smith's deputy street commissioner, the path to treason was more direct and condemned more explicitly by the New York press, particularly after the Civil War.[39] During the summer of 1861, with his mind made up, Lovell surreptitiously made arrangements to move his family south and join the rebellion in September. Lovell's clandestine plans were so effective that no one in New York, save Smith, really knew his intentions. Just days before resigning from the office of deputy street commissioner, Lovell declined Mayor Wood's offer to command the "Mozart Regiment," a contingent of volunteers backed by Wood's new party machine at Mozart Hall (after Wood's secession ploy had failed earlier in the year, he fully supported the Union effort to defeat the Confederacy).[40] The press called Lovell's resignation "disgraceful" and said that "he was not too proud, though a 'Southern gentleman,' to feed at the table and at the expense of those against whom he was conspiring, until concealment was no longer possible."[41] By October, he was a Confederate major general in New Orleans, commanding the Military Department of Louisiana.[42]

Lovell's actions created enough suspicion, though, to spur stories of conspiracy and coercion. On 13 April 1861, he wrote to General McClellan, explaining his and G. W. Smith's position. In that letter, Lovell made a vague appeal, writing, "I trust that mature reflection and close investigation of the whole subject will result in your siding with us."[43] McClellan, of course, pursued a different path that led to commanding Union troops during the war. Lovell's pro-South sentiments, on the other hand, ensured that the traitorous Streets deputy would remain an object of anger and derision in New York. One writer to the *Times* called Lovell a "doubly-dyed traitor" who was "a grand nephew of Benedict Arnold."[44] Another wished Lovell dead from the cannons of the Union gunboats sailing up the Mississippi.[45]

After the war, Lovell attempted to start a rice plantation near Savannah, Georgia, but a hurricane and ensuing flood ruined his first crop in 1867. He made his way back to New York with his family and became an assistant engineer to General Newton.[46] Lovell worked again for the city, initially under Fitz John Porter as city surveyor and later as a civil engineer.[47] As might be expected, Lovell's postwar appointments were not without some controversy. In 1878, a *New York Times* piece questioned, "Have we effected a victory in order to give its fruits to rebels who surrendered important trusts committed by the people to them?" Specifically, the concern was that City Hall was rewarding former traitors with "the first offices of the City."[48] While postwar suspicion of Lovell was understandable, so were the bonds of West Point and the war in general. In Lovell's case, he emphasized his duty in the Mexican War, serving as a president of the Veterans Association of the Mexican War, which provided him a modicum of respectability among his peers in New York.[49]

How Lovell was remembered upon his death illuminates an intriguing aspect of his story. In the *Annual Reunion* of 1884, G. W. Smith wrote the memorial essay for his old colleague and compatriot. Smith took specific care to refute the charges against Lovell's character, insisting that his friend had indeed not "violated his pledged word of honor" to serve the city.[50] What is more, Smith used the memorial tribute to defend Lovell's actions in the fall of New Orleans to the Union Army, a failure that many in the Confederacy blamed solely on Lovell. After the Civil War, the U.S. Military Academy did not allow for the graduates who defected to the South to be memorialized at the school, but there, in the pages of the Association of Graduates' annual publication, was an example of how Confederate veterans could be venerated alongside their Union classmates.[51] Honor was a hallmark value for the veterans of both sides of the war, and the pages of the *Annual Reunion* served as an outlet for West Point alumni to fulfill that sense of honor.

Unlike Lovell's return to Gotham, G. W. Smith's return in 1876 was solely as a businessman. He did not hold any further positions under City Hall. During Reconstruction, Smith remained in Tennessee and Kentucky and re-invented himself as an authority on insurance, writing two books and rising to become the insurance commissioner of Kentucky between 1870 and 1876. For the last decade of his life, Smith lived in New York, where he wrote various articles and papers recalling Mexican War exploits and Civil War battles.[52] Despite the fact that not all was forgiven for the old Confederates in postbellum Gotham, the city was a place of reconciliation and source of redemption for Smith and Lovell. Why? Post–Civil War New York remained a bastion of resistance to Reconstruction and protracted black Americans' struggle for equality in the second half of the nineteenth century.[53]

PORTER AND NEWTON AT THE PUBLIC WORKS

Whereas former Confederates discovered difficulty in securing public positions in New York's Department of Public Works, Union veterans did not, including Fitz John Porter. Nearly a decade before becoming a police commissioner, he first served the city as commissioner of public works. Financially strapped from three years of appealing to various state legislatures to hear his court-martial appeal, Porter succeeded Boss Tweed's man, John Kelly, as commissioner of the department on 1 March 1875.[54] Porter was to be a controversial appointment for Mayor William H. Wickham and City Hall. First, he drew the ire of Tammany men who hated "the sight of epaulettes or the odor of a General."[55] During his first months as head of the Public Works Department, Porter's court-martial appeal was political fodder for Tammany Democrats and the New York papers. Even George Templeton Strong commented that New Yorkers were "a little provoked . . . at [the mayor's] appointing [him] commissioner."[56] Porter's job performance over time countered accusations of cronyism and corruption being made in the press. In an effort to reduce graft and corruption, he reduced pay from $2.00 a day to $1.60 a day for the department's laborers, which enabled him to hire an additional seven hundred workers for the city.[57] During his first round of appointments, Commissioner Porter largely bypassed candidates sponsored by Tammany for those backed by Swallowtail Democrats, which added to the ire of Tammany officials, specifically John Kelly.[58] Porter also benefited from serving as public works commissioner because he could use his actions to demonstrate his innocence. How could a public servant who adhered to the rule of law and fought corruption have been so disobedient on the battlefield? New York gave Porter the chance to prove his critics wrong and garner support for his court-martial appeal.

With respect to the development of Gotham, Porter contributed to the northward expansion of Manhattan as best he could. Recognizing that the path of progress went through Albany, Porter encouraged Mayor Wickham to lobby the state legislature for passage of the 1875 Croton Aqueduct Bill. In Porter's view, older buildings that hindered construction of avenues and boulevards needed to be removed. Additionally, the city government had a responsibility to improve the Croton water system for the people of Gotham.[59] Sometimes Porter went directly to Gov. Samuel Tilden for help in improving city streets and increasing the department's ability to maintain those streets. In May 1875, he asked Tilden to sign the bill eliminating flooded portions of Avenues A and B on the Upper East Side.[60] Where a Tammany appointee might have pursued more funding to make the flooded avenues marginally functional and a potential source of patronage, Porter saw the potential savings to the Public Works' budget. By eliminating the flooded avenues, there would be fewer streets to maintain, and thus more funds to carry out other municipal responsibilities.

Porter may have had the backing of powerful Swallowtail Democrats like August Belmont, but he could not overcome the entrenched power of Tammany Hall, which convinced the board of aldermen not to reappoint him in 1876.[61] Porter's deputy, John H. Chambers, who had been dismissed from the department by G. W. Smith in 1861, was beholden to Tammany leader John Kelly and, as a result, Tammany patronage and graft still flowed from the Department of Public Works. Infamous figures such as Matthew "Rocky" Moore were able to secure almost half of the department's new appointments for Tammany supporters in 1875.[62] When Mayor Wickham and Porter redoubled their efforts to reduce the wages of day laborers in order to hire more for the benefit of the city, the board of aldermen continued to vote them down.[63] Too exhausted to resist the resurging Tammany politicos in 1876, Porter resigned and found employment in New Jersey, running Sen. Theodore Randolph's coal business.[64]

A decade later, another West Point graduate made a try at running New York's Public Works Department. Fresh from the success of clearing the Hell Gate reef formations in the East River, John Newton accepted Mayor William Grace's appointment as commissioner of public works in August 1886.[65] Chosen for his engineering expertise and forty years of leadership experience in the army, General Newton left the military to lead the Public Works Department for an annual salary of $10,000.[66] For two years, Newton attempted to run Gotham's public works with as much transparency as possible. He repeatedly published letters to the editor in the *New York Times* to respond to rumors of wrongdoing within the department. When corruption was proven to have occurred in his department, he would seek to correct or explain how the contract was let under a previous commissioner.[67] Newton worked diligently to extract Tammany's reach from the department's projects. The retired general hired a known reformer, W. A. A. Carsey, to inspect the city's sewers and even secured Fernando Wood's resignation as "keeper of the aqueduct at Croton Lake."[68]

Like those of Wood, William A. A. Carsey's motivations and political agenda were complex, if not contradictory. Carsey, a self-declared activist, supported the Greenback, labor, agrarian, and antimonopolist movements in the 1870s and 1880s. He chaired a Greenback–Labor Party county convention in 1878 and testified before Congress on the merits of regulating the telegraph industry. In 1888, instead of supporting a federal telegraph system, Carsey advocated for "moderate and discreet regulation," sentiments identical to those of the president of Western Union, which happened to be paying Carsey for his support. Two years later, Carsey attended the Farmer's Alliance convention in Florida that created the Populist Party of the 1890s. Carsey's two-faced support illustrates the power of business in this era of nascent reform, and provides context for the environment within which the West Pointers networked.[69] Newton may not have known all of Carsey's affiliations, but he recognized that Carsey was one way to distance himself and the Public Works Department from Tammany.

Newton's effectiveness in reforming the Public Works Department is difficult to assess because the city and party machine politics had changed between Tweed's fall in 1871 and the mid-1880s. After three decades of "bossism," Tammany politics became an accepted way to run municipal government, particularly after the Swallowtail mayors lost power in 1888. Boss Tweed's successor, John Kelly, died on 1 June 1886, making way for Richard Croker. By all accounts, Croker proved to be the epitome of a party boss. He had learned to apply the most effective practices of his predecessors (Wood, Tweed, and Kelly) and, as the head of Tammany from 1886 through 1901, he incorporated Wall Street interests into his schemes.[70] Backed by labor and Irish Americans, Tammany's system of fund-raising and political support became standard practice for party machines, including Thomas Platt's Republican machine. Furthermore, utility companies and big business looked to the bosses to advance their interests.[71] After two years steering Gotham's Public Works Department, Newton had had enough of city politics and became president of the Panama Railroad Company.[72] Newton maintained that position until his death in 1895.

There was no doubt, though, that the public admired John Newton for his accomplishments as a cadet, army officer, and private citizen.[73] Gustavus W. Smith memorialized his classmate in the *Annual Reunion* of 1895. He noted that Newton was born in Virginia and was "an earnest and devout member of the Roman Catholic Church" who had earned an honorary law degree from St. Francis Xavier College.[74] Perhaps there was some irony in the fact that Newton, a Virginian, remained in the U.S. Army throughout his career, unlike his classmates Smith and Lovell, who had experienced success in antebellum New York and still joined the Confederate Army.[75] What is not surprising is that Smith memorialized Newton. Newton's willingness to hire Lovell after the war was a conciliatory gesture that underscored those Military Academy and Mexican-American War connections valued by all three. Moreover, the postbellum reconciliation among them was more eagerly reached in the accommodating political climate created by New York's Democrats and by Gotham's overall embrace of engineering expertise needed after the war.

POLICE COMMISSIONERS—
GENERALS OF ORDER AND REFORM?

The West Pointers' engineering expertise had more practical application for managing the Public Works Department than the city's Police Department. While both departments benefited from the military's experience with organizing and managing large groups, the position of police commissioner required a capacity to lead and organize more than 2,100 officers across thirty precincts.[76] After 1870, the Police Board normally had three or four commissioners, which further complicated management of the Police Department. For much of the Victorian era, controlling New York's police force remained a challenge for municipal leaders. Beyond the actual management of a bureaucratic entity, Gotham's police commissioners had to meet expectations of both city leaders and the populace. Often, those interests were contradictory and required the superintendent of police to take into account political concerns when making practical decisions. Moreover, the city's populace was so diverse that any police action was likely to offend one interest group in the process of placating another. Central authority over Gotham's police was the desired goal of the Police Board, but the reality was that effective department leadership could only come through managing territorial precincts.[77] As with public works commissioners, police commissioners needed to have the capacity to master both the political and practical challenges of the job.

The history of the New York Police Department was a microcosm of the political struggle between city Democrats and state Republicans punctuated by the multiple "charter reforms" decreed from Albany during the nineteenth century. In antebellum New York, Fernando Wood pursued control of the police as a means to secure his base of power. Between 1853 and 1857, Democrats enacted laws that put police commissioners directly under the mayor in the name of police reform, an arrangement generally accepted by most.[78] By the mayoral election of 1856, however, Wood's management of the department appeared to be just another example of his heavy-handed corruption and "infamy," as he coerced councilmen to select his choices for police commissioners. Wood's critics could also point to the contest between the Metropolitans and the Municipals over the summer of 1857 when New York's gangs rioted at will.[79] The state legislature created the Metropolitan police force in the 1857 charter as a means to remove Wood's control over the police. In response, the mayor established the Municipal police force as the city's (really Wood's) law enforcement. City gangs such as the Dead Rabbits, the Plug Uglies, the Five Points gang, and the Roach Guards aligned with one or the other of the two police organizations. As gang warfare raged in the Bowery district, Wood's Municipals and the state's Metropolitans permitted each gang alliance to fight out a proxy war.[80] Horrified by the bloody melee of gang warfare in Five Points and the Bowery, New Yorkers of all classes clamored for the police to establish order—a demand repeated over the next several decades.[81] By 1860, the state-controlled

Metropolitan Police had overcome the Municipal Police, absorbing many of the "Roundsmen" and patrolmen into their ranks.

Given the population growth, as well as the increase in crime and violence that plagued Gotham during the Civil War era, the state legislature might have taken on more than it had anticipated when it wrested control of the municipal police force away from City Hall. To police the city in the 1860s, New York State spent over $10 million, with an annual budget of $1.7 million increasing to $2.8 million by 1869. In 1863, the state built a police headquarters building at 300 Mulberry Street, and then constructed an additional wing in 1869.[82] The upsurge in wartime prostitution, gambling, and alcohol consumption added to the stress of the police leadership, as Mayor Oakey (a Tammany mayor) and early reformers demanded that the Metropolitans stem the suffering and decadence they saw in the city. Tammany "reform" equated to defending civil rights and standing for "home rule." When the police took action, violence escalated across the precincts. Gangs and criminals attacked patrolmen on duty, and there was a corresponding rise in incidents of police brutality. Compounding the situation was an ineffective court system that tended to grant lenient bail to arrested suspects and not support the actions of the police. By 1870, the state-reformed Police Department appeared no better than the one managed by the city in 1857. Perhaps undermanned and physically too far removed to control the department, Republicans in Albany approved a new charter proposed by Tweed that restored authority over Gotham's police to a locally controlled board.[83]

The Tweed charter of 1870 put control of the Police Department under a board of four commissioners, each appointed for a term of different length. To ensure that the mayor not have the single point of control over the police, the charter mandated that the board report to the Common Council and granted the police superintendent the most direct control over the force.[84] Tammany could steer the Common Council and the commissioners as it desired. In 1873, after Tweed's fall, another reform committee issued a new charter with a new three-man board and granted even less authority for the mayor to control the police directly. Now, under the 1873 charter, the mayor could manage the Police Department only through the Police Board, whose members he appointed. The superintendent answered to the Police Board and not the mayor. As result, the superintendent had primary authority over the police.[85] This system appeared to work until Henry Smith, a member of the Police Board, died in 1874. Neither the board of aldermen nor Mayor William Havemeyer could agree upon an acceptable replacement. During the interim period, two other police commissioners crossed Tammany leader John Kelly by not appointing some 3,136 election inspectors (this function fell under the Police Department in 1874) designated by Tammany. Kelly had the commissioners indicted. In response to Kelly's attacks, they resigned, and when Mayor Havemeyer tried to reappoint them, the press cried foul. As the scandal came to light, Mayor Havemeyer died, leaving the public to demand reform from City Hall and the Police Department throughout the winter of 1874–1875.[86]

In response to the rampant corruption that they perceived running through the department and City Hall, Mayor Wickham and Police Commissioner John Voorhis took drastic actions to reform New York's police at the end of April 1875. First, the Police Board abolished three precincts and transferred their nineteen officers to other precincts around the city. Then, it reprimanded and transferred another twenty-three officers to patrol duty.[87] Finally, as the *New York Times* reported, Mayor Wickham undermined Tammany's authority by submitting the names of Swallowtail Democrats to the board of aldermen as candidates to head the Park Commission and the Police Department. He proposed that August Belmont go to the Park Commission and Gen. William F. Smith to the Police Board. (William F. Smith was not related to Gustavus W. Smith, who had defected to the Confederacy back in 1862.) The aldermen did not approve Belmont's nomination, but William F. Smith's nomination was not objectionable to Tammany, as John Kelly reportedly approved General Smith's appointment.[88] It was in this context that Smith entered the fray of Gotham's municipal government.

"Baldy" Smith, as he was known to his peers at the U.S. Military Academy, graduated fourth of forty-one cadets in the class of 1845. The son of a Vermont farmer, he excelled at engineering as a senior cadet. Upon his commissioning in the Corps of Topographical Engineers, Smith served on topographical survey expeditions in the Great Lakes. He came back to West Point to teach mathematics, where he remained through the Mexican War. As a captain, he became engineer secretary of the Lighthouse Board and supervised construction of lighthouses at Montauk Point, New York; Sandy Hook, New Jersey; and Cape Canaveral, Florida. Smith suffered bouts of malaria-induced depression as a result of contracting the disease while on a survey expedition in Texas.[89]

During the Civil War, Smith remained in the Union Army, initially leading the 3rd Vermont Volunteers in northern Virginia, and later rose to the rank of major general, commanding a division under McClellan in the Army of the

Potomac. Similar to Fitz John Porter, Smith was openly critical of McClellan's successors. In Smith's case, he signed a letter to the president criticizing Gen. Ambrose Burnside's performance after the battle of Fredericksburg, which had been fought in late 1862. Later, as a corps commander, General Smith criticized Gen. George G. Meade for the high number of Union casualties suffered during the Battle of Cold Harbor in June 1864. After failing to lead his corps through the Confederate picket lines at Petersburg, Virginia, Smith was relieved at the behest of Gen. Benjamin Butler. Loyal to McClellan in battle and a Democrat to boot, Smith did not appear to have a promising postwar future if he were to stay in uniform. Still, when he resigned from the army in 1867, he had built an impressive record leading Union troops in battle.[90]

After the war, Smith became president of the International Ocean Telegraph Company, where he negotiated with Spain for the construction and management of a telegraph line between Jacksonville, Florida, and Havana, Cuba. In 1873, he sold the company to the Western Union Telegraph Company and toured Europe with his family for two years.[91] He returned to New York from Britain in January 1875, arriving in the middle of Gotham's police corruption uproar.[92]

As was to be expected, Smith assumed his new position planning to implement a program of reform over the Police Department. During a Police Board meeting in August 1875, he proposed plans that would change the rules of the board and curtail the corruption that he thought originated with his fellow members. Smith, not one to hold back his opinion, called Commissioner George Matsell's patron, the *National Police Gazette*, "a school for thieves" that was "the best text-book for young villains."[93] Matsell was a veteran of police affairs dating back to the founding of the New York Police Department in 1845, when he was the first superintendent.[94] Clearly, Smith was a challenge to the status quo. A few weeks later, when the Police Commission was investigated for allowing the police to protect illegal gambling houses, the investigator specifically excluded Smith from the charges because of the general's "distinguished character as a soldier and citizen" and "his plainly evident distaste of his associates and their acts."[95] Just as he had done during the Civil War, General Smith openly criticized his superiors and peers when he thought it justified, and he received positive press for his forthrightness.

However, as he had found in the army, there were limits to how outspoken or critical one could be. At two points during his tenure on the Police Commission he almost left the board. In October 1875, when the allegations of corruption compelled Mayor Wickham to investigate the board, Smith and fellow board member John R. Voorhis submitted resignations. Both commissioners knew that the mayor would not accept their resignations, but their actions were part of a larger scheme to remove Matsell and another commissioner, Abraham Disbecker, who were the mayor's real targets. By the end of the year, Wickham had succeeded in removing Matsell and Disbecker, creating great expectations for Smith to reform the department. As an outsider, Smith could not really change decades of practices and relationships, especially by being a vocal critic of the very organization to which he belonged.

Five years later, Smith almost left the board again when Mayor Edward Cooper sought to put "Tildenite" Democrats on the Police Commission. As a Swallowtail Democrat, Cooper wanted to expunge any Tammany or Republican commissioner from the Police Board and replace him with a non-Tammany Democrat. Smith, who by then was president of the Police Board, appeared to be too close to Tammany's John Kelly in Cooper's estimation.[96] In March 1879, Cooper sent Smith a letter blaming the West Pointer for allowing "the streets to remain in a foul state, and the efficiency and discipline of the Police force to deteriorate." Cooper charged that Smith's "personal dissensions" prevented him from "co-operating with his fellow commissioners" and, as a result, caused the overall failure of the Police Department.[97] What was viewed as bold and courageous in 1875 now was seen as obstructionist and uncooperative. Enemies on all fronts were using Smith's candid nature against him. By the summer of 1880, Smith's lawyers, Elihu Root and Willard Bartlett, successfully defended Smith's position that Cooper's actions were illegal, and the state supreme court allowed Smith to remain president of the Police Board through the end of his six-year appointment.[98]

Because of the many controversies that touched the Police Board between 1875 and 1881, William F. Smith was not effective at reforming a system entrenched in boss politics. At times he appeared to uphold the status quo of favoritism and patronage in the New York Police Department, condoning the practice of selling off captain's appointments and promoting war veterans. He resisted efforts to combine the Board of Health with the Police Department.[99] Smith missed many Police Commission and Health Commission meetings toward the end of his tenure.[100] At other times he could be the reformer working within the current practices of the department, ordering Sunday police raids to enforce saloon laws on businesses illegally selling beer and liquor.[101] General Smith could also be partial to U.S. Army veterans of the Civil War. According to Capt. Timothy J. Creeden's testimony before the Lexow Committee in 1894, Creeden's army service "paved the way for his first two promotions" in the Police Department.[102] In contrast, Smith did not think highly of police detectives and lectured them several times on their duties and responsibilities.[103]

Like other police commissioners, Smith was subject to scrutiny and public judgment, even for the slightest misstep. A *New York Times* reporter noted that carriages waiting outside Smith's house during a dinner party one evening failed to light their lamps in compliance with the municipal ordinance.[104] Allegations of indiscretion, drunkenness, and dereliction of board responsibilities were common during Smith's time on the Police Board.[105] When Smith resigned from the board in 1881, he recognized the difficulties that he may have caused for his fellow commissioners. Delivering his letter before the board, he announced that he had news they would "relish."[106] After quickly collecting papers from his office, Smith left 300 Mulberry Street and the scrutiny of public office.[107]

Upon completing his term as police commissioner, General Smith returned to his first calling: engineering. Through his connection with General Newton, who was army chief of engineers in 1881, Smith became a civil engineer contracted to complete projects in Delaware and along the eastern shore of Chesapeake Bay. Eventually, Smith retired in Philadelphia, where he died in 1903.[108]

It was three years before another West Pointer would serve on the Police Board. In October 1884, Mayor Edson asked Fitz John Porter to serve out the term of a recently deceased police commissioner. Similar to the time Mayor Wickham tapped Porter after the death of the public works commissioner in 1875, Edson wanted a strong anti-Tammany Democrat in the open position. As described earlier, Porter's experience in the war and his ability to stand up to machine coercion made him a favorable choice to put on the Police Board. During the summer, President Chester Arthur had vetoed the 1884 bill exonerating Porter, and the beleaguered general found himself on the brink of financial ruin (two years earlier the president had succumbed to political pressure and approved an 1882 bill reinstating Porter's citizenship and rank of colonel, but stopped short of exonerating Porter's conviction).[109] Even though the city leaders knew Porter from his time as commissioner of public works, his appointment to the Police Board caught the other commissioners completely by surprise.[110] He was a bit of a Democratic *cause célèbre* because Congressman John A. Logan (leader of the Grand Army of the Republic, or G.A.R., from 1868 to 1871) and other congressional Republicans had denied his appeal for so many years. Appointing Porter to the Police Board was one way to show the Republicans that they did not control New York. Others painted General Porter as a Tammany pawn who previously had catered to John Kelly's patronage system.[111] Shortly after Porter's appointment, Kelly and a delegation from Tammany Hall did pay an office call to the new police commissioner to wish him well, but Porter remained a non-Tammany Democrat.[112] Whether or not Porter's associations with Kelly were real or mere perception did not matter. Porter was the mayor's choice, and as such, the mayor successfully lobbied to keep Porter on the Police Board in January 1885.

Meanwhile, at his office in police headquarters, Porter spent as much time preparing his court-martial appeal as he did attending to department matters. On "Commissioner's Office" stationery, he copied by hand hundreds of letters and documents from 1862 to send to various veterans, lawyers, and politicians working the case.[113] With the 1884 election of President Grover Cleveland, the first elected Democratic chief executive since Buchanan, Porter's hopes increased that he could get an appeal through Congress and signed by the chief executive. Porter had to act quickly in putting together another effort to clear his name, and therefore, spent less time on police matters, even skipping official functions like the annual police captain's dinner at Delmonico's.[114]

Porter's case commanded the attention of the press, politicians, and veterans. The G.A.R. lobbied hard to stop Congress from passing Porter's relief bill. G.A.R. commanderies across the country passed resolutions requesting that Congress not restore General Porter to the army.[115] Partisan politics escalated in January 1885 as Democrats exploited a Republican bill to add U. S. Grant by name to the retired roll. Sponsored by Vermont senator George F. Edmunds, the Edmunds bill first proposed to give Grant a full general officer's pension in 1881. The urgency to pass the measure increased in 1884 when Grant went bankrupt. If President Arthur were to sign the bill giving Grant his pension, then he would contradict his earlier objections when vetoing the 1884 bill exonerating Fitz John Porter and allowing an earlier pension bill for Grant.[116] In a ploy described as "contemptible" in the *New York Times,* Henry W. Slocum led House Democrats in challenging the Grant pension legislation as a way to embarrass President Arthur for vetoing the Porter bill in 1884 and for vowing not to sign any bill that mentioned Grant by name.[117] At the time, Slocum represented Brooklyn in the House of Representatives. Like Porter, he had commanded a division under General McClellan during the war, and he was sympathetic to Porter's cause. Slocum's arguments only delayed the inevitable.

President Arthur's final act as chief executive was signing the Edmunds bill that afforded General Grant, by name, a full pension.[118]

Determined to the end, Porter drove on with his appeal while tending to police matters, but the Police Department was not his priority. When the Police Board nominated Porter to be its president, he declined the nomination.[119] With Democrats in the White House and in control of the House of Representatives, Porter knew that this was his best chance to get an appeal through, and so he devoted his energy to capitalize on a potentially limited window of opportunity.[120]

Also at this time, New York had a new mayor, William Grace, another anti-Tammany Democrat. Shortly after taking office, Mayor Grace approached Porter about switching from police commissioner to being head of the Fire Department. Porter's political allegiance was too unknown for Grace, and the mayor thought the anti-Tammany portion of the Democratic Party support would be more secure by replacing Porter on the board. Mayor Grace's discomfort was understandable. In calls for further investigation of Tammany corruption, lawmakers in Albany named Porter as someone "warmly in sympathy with Tammany."[121] At the same time, Porter introduced reform resolutions before the Police Board that were to abate patrolman activities not germane to law enforcement, such as the practice of policemen selling tickets to political fund-raisers.[122] Additionally, as a police commissioner he sat on boards of inquiry that investigated police misconduct. In spite of the mayor's concerns, Porter remained a police commissioner through the end of Grace's term.

In time, Porter did make a change and became the Fire Department commissioner in 1888. Mayor Abram Hewitt convinced Porter to take the post vacated by Richard Croker, who was securing his place atop the Tammany machine.[123] But Hewitt, the last of the "reform" Swallowtail mayors, could not outmuscle Croker for control of Gotham's officials.[124] With this reemergence of Tammany's sway over City Hall, General Porter realized that he was being marginalized and pushed aside, and resigned as fire commissioner a year later.

Perhaps Porter's daughter gave the best assessment of West Pointers who served as police commissioners in New York. She told a Porter biographer that "Father made a good record in the Police Department, but not a spectacular one."[125] She might have made a similar judgment of William F. Smith. The record of West Pointers as head of the New York Police Department is mixed. Certainly they tried to reform the organization responsible for enforcing the law across the city, but Tammany remained more powerful than W. F. Smith,

Fitz John Porter, and the Swallowtail Democrats. The West Pointers attempted to manage the department honestly and in the public interest. However, Tammany appointees and decades of institutional corruption were too pervasive for one or two individuals to stem, especially since each was one among a committee of four. When compared to changes achieved by the Lexow Committee and the appointment of Theodore Roosevelt as president of the Police Board, Smith's and Porter's efforts appear as failures to reform the Police Department during the 1870s and 1880s.[126] To be sure, the Military Academy men were not Progressive reformers in the spirit of Theodore Roosevelt and Jacob Riis, but their experiences do represent preliminary efforts at reform in postbellum New York.

Smith and Porter possessed the expertise and experience that the city's Democrats hoped would improve policing of the growing metropolis. The Democrats expected them to be loyal appointees because both men had been vocal supporters of General McClellan during the Civil War. Moreover, they were true McClellanites since each general's military career had been marginalized because of his loyalty to the "Young Napoleon."[127] Another key consideration was that Tammany failed to draw Smith and Porter under its control. So, for the anti-Tammany Democratic mayors, the West Pointers were model candidates to put on the Police Board.

How did Smith and Porter benefit by accepting these posts? By serving as police commissioners they received income, status, and a place in the public's memory of the Civil War. First, the financial compensation of $5,000 per year made for a very comfortable quality of life, particularly for Porter.[128] Second, the job provided them a status that they viewed as commensurate with their education and experience. In addition to veterans' events and annual reunions at West Point, Smith and Porter attended dinner parties, weddings, and funerals of New York's social elite.[129] Smith fared better than Porter with his business interests and was able to live in Manhattan after the war. Thus, serving as police commissioner reinforced his social status. Porter, on the other hand, lacked income to move his family to the city during his years as commissioner and lived in Morristown, New Jersey.[130] In spite of his financial challenges, Porter's status remained high among Democratic New Yorkers because of his public struggle to correct an injustice inflicted by Radical Republicans. Finally, the police commissioner position provided these two veterans a means to bring attention to their service in the Civil War. A reference to what they did during the war or the title of "general" usually accompanied any mention of their name in the press.[131] Thus, their military service contributed to their sense of identity as public officials, as well as to how others perceived them.

Still, why did Smith and Porter fall short of the reform goals expected of them and that they expected of themselves? Merely saying that they were too weak to counter the party machine at Tammany does not get to the essence of the question. Arguably, the same characteristics that made Smith and Porter desirable nominees for Democratic appointments at City Hall also limited their potential to make reforms. Experience and expertise gained from a military background worked both ways. From their cadet experiences at the Military Academy through their military service in the Civil War, they learned that those who conform best to the norms of the organization often rise to the top. By design, nineteenth-century militaries required discipline and structure to master the chaos created by war. Those officers who studied the hardest and accumulated the most experience were likely to possess the best expertise. The military system favored conservatism, especially if one was not a military genius.[132] Discipline and order were ideals imbued in these military officers from their first day as cadets and every post thereafter. Obedience to army superiors and rules was necessary for winning battles. Deviating from military order and discipline could lead to failure.[133]

In this context, leading an army unit or leading the New York Police Department did not require substantial change. It was more a matter of imposing order over an organization failing to conform to rules and regulations than it was of creating a new organization. For the commissioners who came before the Progressive era, reforming the city's police force was a task to be done within the existing system of government, political relationships, and, perhaps most important, the interests of Gotham's leaders.

Radical change would come to the Police Department a decade later with Theodore Roosevelt and Jacob Riis, when Mayor William L. Strong, a businessman who represented the interests of other merchants and professionals, demanded "honest, efficient, and businesslike" municipal government.[134] Ironically, Roosevelt took a more military approach to policing than his West Point predecessors. He thought that managing New York's police force required "many of the principles" that one could "obtain in the army."[135] Under Roosevelt, the New York Police Department's purpose was to wage "[war] on dishonesty" and "increase efficacy."[136] He and Commissioner Avery Andrews changed the uniforms and reorganized the department like a military organization. Additionally, Roosevelt began the practice of publicly recognizing policemen for feats of heroism, rewarding them with promotions.[137] Granted, as police commissioner, he was more known for prowling the night streets with Riis than any other aspect of the job, but his adaptation of military organization and personnel incentives had a lasting effect on Gotham's police.[138] Thus, eventual Progressive reform of the department occurred through the use of military practices, not military veterans.

The other postwar development associated with Civil War veterans was the creation of the G.A.R. by Benjamin Franklin Stephenson, a veteran Union Army surgeon. Founded in 1866 under the pretense that Civil War veterans missed the camaraderie of the field camps, the G.A.R. was formed as a brotherhood of veterans who could also support Illinois Republicans John Logan and Gov. Richard Oglesby.[139] For the first decade of its existence, the G.A.R. was both a veterans' organization and a vehicle to raise support for the Radical Republicans. In 1868, the G.A.R. boasted some 250,000 members led by Logan and the New York commander, Gen. Daniel Sickles. Despite the G.A.R.'s outward inclination to support Radical Republicans in elections, New York led the G.A.R.'s eastern departments in membership through 1868. Still only three years removed from the war, Union veterans were Unionists first and political party supporters second.

By 1871, at the Fifth Encampment, Maj. Gen. Ambrose Burnside—who was also a former Republican governor, of Rhode Island—set a policy that denounced any member who used his G.A.R. post for political advantage. As George Lankevich has shown, the veterans' group almost became defunct in the mid-1870s, revived as a fraternal order by 1879, and in the 1890s was "a powerful lobby for pensions, 'correct' history, and a particular brand of American nationalism."[140] Veterans of both parties supported the G.A.R. pension agenda during the 1880s, and the organization proved to be one of the "most powerful political [lobbies] of the age."[141]

Other than Grover Cleveland, every postwar president through 1900 was a member of the G.A.R. However, by 1895, the group had evolved into an organization dedicated to patriotism across the nation. Members, including Viele, campaigned for "a comprehensive plan of patriotic education."[142] Contrary to the wishes of Burnside in 1871, the G.A.R. remained a politically active organization, although it was more an issue-oriented group than it was an auxiliary of any political party. At the end of the century, the G.A.R. was also part of a national movement to embrace military education values.[143] Veterans may have distanced their actions from either party, but they felt obligated to make speeches urging education boards to adopt "military rules" and discipline in the instruction of school-age boys.

As Americans reflected upon the nineteenth century, West Point and the actions of its graduates were treated as key ingredients for the successes of the United States to that point. Some, especially veteran military and naval officers, viewed the Academy's military program and the professionalism of the American military officer corps as necessary elements of any future national prosperity.[144]

EMBRACING MILITARY GENERALS AND POSTWAR CIVIL LEADERSHIP

In New York, the West Point veterans who served City Hall were evidence of a national affinity toward military experience in nineteenth-century America. The willingness of city leaders to recruit and appoint individuals who studied engineering at West Point and fought in the Civil War was not unique. Every community in the nation had come to accept the new realities created by the war. For example, every section of the nation experienced extensive loss through the pervasive carnage of the war (with the South suffering the most). Drew Gilpin Faust notes that "shared suffering" overrode regional differences and provided the foundation of reunification.[145] Along the same line, the country's common experience with the two armies led to a recognition that the expertise and attributes of military leaders were transferable to civilian positions, both elected and appointed. Moreover, military personalities were familiar to both political parties after the Civil War.[146] Democrats could depend on veterans voting in the urban Northeast and the Midwest, not to mention former Confederate officers in the South. For the Republicans, the G.A.R. proved to be a powerful organization not only for election campaigns, but also for lobbying special-interest legislation like the veterans' pension bill during the 1870s.[147] Without a doubt, the issues that drove American voting behavior during the periods of Reconstruction and after were not identical across the nation. For instance, black codes in the South shaped the outcome of postwar elections there, while new immigrants in the industrializing cities often determined political outcomes in the North. However, the shared experience of the Civil War cut across all issues and voter concerns, with the election of 1868 being a key example.

Leading up to the 1868 presidential election, there clearly was a connection between military leadership during the war and expectations for national political leadership afterward. Although Republicans vacillated on the issue of black suffrage just three years after Appomattox, they still supported Radical Reconstruction as the method to bring the South back into the Union. Ulysses S. Grant, who both had proven his commitment to the abolitionists' cause during the war and was tepid on racial equality, became the acceptable choice to lead Reconstruction to a suitable completion.[148] Unlike General McClellan or Henry Halleck, who had ridiculed Lincoln's war to end slavery, Grant understood the desired objective of the war and pursued a determined, sanguine strategy to achieve it.

Grant was the first of five Republican presidents who had been general officers during the war. The others—Rutherford B. Hayes, James Garfield, Chester Arthur, and Benjamin Harrison—were "political generals" appointed at the behest of various state leaders and constituencies, but all had worn the uniform as generals fighting for the Union cause.[149] For them, military service was an unwritten requirement to demonstrate loyalty and patriotism in the process of touting their qualifications to lead the country.

After Grant's two terms, Republican presidents shifted away from "ideological politics," where one's stance on race and Reconstruction issues mattered the most. During the election of 1876 and the Compromise of 1877, President Hayes concentrated on "organizational politics," where one's ability to build support and loyalty determined election outcomes.[150] The momentary emergence of Liberal Republicans in 1872 also contributed to the end of Radical Republican ideals dominating the party's agenda. Guided by Carl Schurz, disenchanted Republicans briefly sought a third party more focused on anticorruption and reform.[151] As a lack of common ideology diffused political affinities, organizational power and control became more important in presidential politics.

Perhaps the most insightful contemporary commentary on the presidency of the age came from British historian James Lord Bryce's two-volume study of the United States, *The American Commonwealth*. In a chapter titled "Why Great Men Are Not Chosen," Lord Bryce noted that "the President need not be a man of brilliant intellectual gifts." Instead, he compared the work of the American president with that of a business executive or "the manager of a railway" who served more as supervisor of employees and maker of practical decisions.[152] The other keen insight from Bryce's book was that regardless of a candidate's "commanding character" and "perfectly clean record," he could not run for the presidency unless he "was in the war."[153] In effect, Civil War service was a prerequisite for any serious presidential candidate, and moreover, the more senior the military rank reached during the war, the higher the expectations were that the veteran could manage the national government.[154]

CONCLUSION

Republicans tended to dominate postbellum politics nationally, while Democrats maintained strongholds in Gotham and the Old South. What becomes clear in this composite of local and national politics is that one's military service during the war proved to be more than just something to commemorate on Memorial Day and the anniversaries of battles. For the national Republicans, military service was a sign of commitment and patriotism. In New York, Democrats saw military service as evidence of expertise gained from real-world experiences. Whether a park commissioner, commissioner of public works, or head of the Police Department, the West Point graduate possessed the credentials not only of a wartime leader but also of a Military Academy graduate. Party allegiance and connections mattered in Gotham, but engineering expertise allowed for greater flexibility in the level of commitment to party leadership and machine bosses. Perhaps Bryce's observations about chief executives also apply for the commissioners in Gotham's municipal departments. "Brilliant intellectual gifts" were not required to manage the Public Works Department or sit on the Police Board. Instead, those positions needed practical managers who could run the city like a railroad. The records of Fitz John Porter, William F. Smith, John Newton, and even Gustavus W. Smith and Mansfield Lovell all reflect such qualifications. Furthermore, Gotham insulated these veterans from the derogatory effects of postbellum Radicals such as the G.A.R. and John Logan, who singled out West Pointers as "the scourge of the regular Army."[155]

Thus, the connection between the city and veterans in municipal service was reasonable, given the veterans' circumstances and the needs of city leaders. As Civil War veterans in the prime of their professional lives, they desired a place to maintain, increase, and, in some cases, mend their status. City Hall needed their expertise and reputations to preserve Democratic power with and without Tammany Hall.

Later, in other cities like Staunton, Virginia, Progressive reform led to the innovation of the "city manager," a professional official appointed by elected commissioners or city councils. Commissioners charged city managers with the task of mitigating the corruption and graft of city politics by applying proven business models popular at the turn of the twentieth century. Businessmen recognized that commerce and industry required an efficiently run infrastructure of paved roads, utilities, rail depots, and port facilities.[156] Many of these first city managers in the first decade of the twentieth century were engineers. However, New York did not adopt the city manager concept, because authority granted through successive city charters proved more than sufficient to meet the needs of its citizens. The Public Works Department as well as other city commissions held the power to build sewers, waterworks, and roads throughout Manhattan and beyond.[157]

While the most apparent transformation of New York came in the 1890s, both the West Pointers and city leaders achieved their goals in the 1870s and 1880s. Gotham remained mainly in Democratic hands, and the former generals redeemed their professional status and military legacy. In finding their redemption in postwar New York, Military Academy veterans proved to be bridging figures between the antebellum city and the metropolis that the Progressives would eventually seek to reform.

8

THE EMERGENCE OF MODERN AMERICA
IN WEST POINT'S NEW YORK

Your duty here at West Point has been to fit men to do well in war. But it is a noteworthy fact that you also have fitted them to do singularly well in peace. The highest positions in the land have been held, not exceptionally, but again and again by West Pointers.

—THEODORE ROOSEVELT, address commemorating the centennial
of the U.S. Military Academy, 11 June 1902

During the commencement week activities of 1902, the United States Military Academy celebrated its centennial anniversary. In addition to the usual pomp, ceremony, and rituals associated with graduation and commissioning, there was a parade of distinguished visitors highlighted by President Theodore Roosevelt. Most of West Point's Civil War veterans had passed away by 1902, but there were still 154 alumni living, among some 2,064 antebellum graduates. Of the graduates who had participated in New York's rise, only William F. Smith, the former police commissioner, remained. Egbert Viele had passed away just eight weeks prior to the centennial celebration.[1] In attendance were alumni and veterans of the recently fought Spanish-American War, as well as congressmen, diplomats, university presidents, and an assortment of veterans' organizations. On that rainy June afternoon, the Military Academy not only commissioned another fifty-four graduates, it also paid homage to the contributions of the institution and its graduates over the previous century.

Roosevelt's remarks quoted above aptly memorialized the place of the school and its alumni who had served the nation in war and peace. The president praised the Academy graduates, noting that each had "given a greater amount of service to the country through his life than [had] the average graduate of any other institution" in the United States. Roosevelt extolled West Point as "absolutely American" because it selected its student body with a "mathematical exactness" to represent every geographic area of the country and "every walk of life."[2] Given the Military Academy's place in the nation's history, who could have argued otherwise? In the president's view, the Academy had met or exceeded all expectations since its creation in 1802. Moreover, West Point had become an intrinsic component of the nation. In addition to their "devotion upon countless battlefields," the chief orator for the day, Horace Porter, remarked that the West Pointers "attested their usefulness in all the civil walks of life—in science and art, in trade and commerce, in literature and oratory, in theology, law, diplomacy, and statesmanship, from the modest engineer to President of the Republic." The U.S. Military Academy at one hundred was "absolutely American" and "absolutely democratic."[3]

The legacy of the Long Gray Line educated by Thayer's system and Mahan's engineering curriculum was the basis for the centennial celebration observations. In fighting America's wars, building the nation's infrastructure, and expanding the United States westward across the continent, West Pointers were critical leaders and engineers enabling the progress that marked the United States at the turn of the twentieth century. While the president applauded those aspects of the school that illustrated his sense of reform, fairness, opportunity, individualism, masculinity, and "the strenuous life" in the Progressive Era, the Military Academy was not a Progressive institution.[4] West Point in 1902 really was more a reinforcement of older, conservative tendencies than it was an expression of Progressive ideas. At the Military Academy's bicentennial in 2002, historians described nineteenth-century West Point as "a conservative institution that favors continuity over change." Continuity was a virtue that had enabled tradition and routine to produce an officer corps capable of leading the United States in the wars of the nineteenth century.[5] A subsequent survey of the U.S. Military Academy in its second century concluded that "change, not continuity, best describes the history of West Point" since 1902. As Academy leaders struggled to keep pace with—or to be more accurate, to catch up to—the new realities of science, demographics, and engineering education emerging in the early twentieth-century Unites States, West Point still centered cadet development on "character and intellect."[6] The enduring emphasis of the Military Academy on developing character and intellect enabled its nineteenth-century graduates to shape professionalism, urban development, and the United States' memory of its wartime experiences.

Throughout the nineteenth century, professionalism, urban growth, and war produced a set of national values that juxtaposed capitalism with labor, intellectualism with anti-intellectualism, and progressive change with a preservation of the status quo.[7] By the end of the Victorian era, Americans had transformed into a people of mixed ethnic traditions, class, and race: a mosaic of the social and political change informed by the narrative of the nineteenth century, and a narrative that made modern America.[8] The Military Academy with its conservative influence contributed to the emergence of that modern era.

Understanding how West Point's legacy informed the rise of modern America becomes much clearer when seen through the lens of the graduates who came to New York City between 1817 and the 1890s. While not among the Progressives of that era, Academy alumni in the nineteenth century were antecedents to the Progressivism that occurred in modern America. The Military Academy produced secondary yet important actors in Gotham's rise during the Victorian period. In this capacity, West Point graduates helped create a genuine identity that civilian New Yorkers aspired to, respected, and assimilated. The cadets taught under Thayer and Mahan's engineering curriculum were part of the city whose population included the Knickerbocker elite, businessmen and speculators, professionals, artisans and laborers, nativists, and an ever-increasing number of immigrants. New Yorkers were easily the most diverse of all Americans at the end of the nineteenth century. Across all five boroughs were people of every race, color, creed, and socioeconomic group in the United States.[9] Without the municipal infrastructure to connect the boroughs, there would not have been a Greater New York to draw the millions who came to identify themselves as New Yorkers. Engineers, heavily influenced by former students of Thayer and Mahan, literally bound the city together and became a recognized body having expertise for making New York function.

Notably, though, as conservative Democrats, these West Point graduates fell short in improving circumstances for millions of African Americans in the city. The boroughs may have been connected physically, and the municipal departments functioned well enough, but mainly for white New York at the turn of the twentieth century. Only in the Progressive Era would prospects for African Americans, women, and new immigrants have a chance to improve.

Beginning with the antebellum desire to bring water to the city, Gotham's demand, and the demands for engineering expertise of other cities such as Boston and Philadelphia, drove the expansion of engineering education and technology. As explained in chapter 4, engineers who trained under the civilian master-apprentice system, veterans of the Erie Canal construction, and those educated at West Point all came to the city in the 1840s and 1850s. The confluence of these three engineering traditions in New York led to the creation of the ASCE in 1852. Without the Military Academy, the ASCE would have been deprived of a key constituency during its formative years. West Point–trained engineers were predisposed to using French civil engineering principles because of the influence of the Ecole Polytechnique. Thayer's and Mahan's trips to post-Napoleonic France firmly rooted the Military Academy's courses in the French school's example. The Military Academy's adoption of French ideas was not the only instance of Americans exploiting French examples in that era, but it was a momentous precedent. Thayer's insistence that West Point model the French engineering institution reinforced the domination of French thought in American math and engineering education during the nineteenth century.

In New York the nexus between military professionalism and the rise of engineering professionals became obvious. The Survey Act of 1824 ensured that the nation as a whole would look to West Point engineers to design and build the infrastructure that would increase commerce and trade. Additionally, the Mexican-American War validated the efficacy of having West Point–trained officers serve in the army. Winfield Scott's remarkable campaign of 1847 was due in large part to the leadership and actions of young lieutenants and captains from the Military Academy.[10] Thus, in war and peace, West Point produced officers who provided indispensable engineering expertise. Viele and George Greene demonstrated to city leaders that engineering expertise and army experience were desirable prerequisites for men heading large public works projects. And conversely, military officers realized the opportunity for financial success and increased status in meeting the demands of New York's development. Moreover, since the army's Corps of Engineers only accepted the top two or three graduates of each class, other Military Academy alumni who wanted to pursue their engineering interests had to leave the army to do so. Gotham presented multiple opportunities for them to apply their engineering expertise as civilians. Viele, for example, made the most of his topo-graphical engineering know-how to impress Fernando Wood in 1855. George Greene acted similarly, finding work on the Croton Aqueduct under Alfred Craven. After the Mexican-American War, the city and Academy engineers both benefited from these types of relationships. New York had skilled engi-neers constructing its latest projects while West Point alumni improved their reputations as civilian engineers. In antebellum New York, the employment of former cadets in municipal projects was a validation of the Military Academy's *raison d'être.*

The compensation awarded for engineering expertise also helped entice West Pointers to New York. Considering that the average tailor in mid-nineteenth-century New York earned between $300 and $500 per year, and a master crafts-man cloth cutter could make as much as $1,500 per year, engineering posi-tions paid substantially higher than working-class jobs.[11] Antebellum salaries of $2,500 for Viele as engineer-in-chief of Central Park, $5,000 for David B. Douglass as engineer-in-chief of the Croton Aqueduct, and $2,000 for Greene as an engineer in the Croton Aqueduct Department rewarded these engineers for their expertise and helped them maintain their high status in the city.[12] After the Civil War, engineers' salaries remained competitive with those in other professions, and continued to reward expertise. George McClellan's annual earnings more than doubled from $6,000 in 1868 to $13,000 in 1875. Granted McClellan received income from his business endeavors that included the Stevens Battery and several railroads, but his engineering background and leadership experience made him more than qualified to head the city's Docks Department in the 1870s. In 1866, Greene's Croton salary increased to $4,000.[13] By 1886, New York was paying $10,000 per year to the head of the Department of Public Works, retired general John Newton. While the Croton pay was Greene's main source of income after the war, both McClellan and Newton needed their salaries to supplement their other incomes in order to maintain the quality of life to which they had grown accustomed.[14]

Individual alumni and New York's municipal departments were not the only parties to benefit from this relationship. The U.S. Military Academy also experienced improvements in its institutional reputation and status from the graduates serving as engineers in the city. Over 150 graduates practiced civil engineering in the 1840s and 1850s, proving the peacetime merits of the Mil-itary Academy, especially when the school's critics questioned its value to the nation. Canal construction, the Baltimore and Ohio Railroad, and the Croton

Aqueduct were tangible outcomes of the engineering education grounded in Thayer's curriculum.[15] West Point's reputation as the "First Engineer School" helped its graduates to secure teaching posts at new engineering programs emerging at American colleges from the middle to the end of the nineteenth century. By 1886, the ASCE readily acknowledged the significance of the Military Academy as one of the top three sources of engineering expertise in the United States.[16]

At the same time, the Military Academy contributed to a sense of professionalism in the army. As William Skelton notes, "The proportion of graduates in the [army] officer corps rose dramatically: from 15 percent [of all officers] in 1817, to 64 percent in 1830, and to 76 percent in 1860 (exclusive of paymaster and medical officers)."[17] As a result, "West Point helped shape the relationship between the Army and the civilian world. On one hand, the special circumstances of the Academy—its isolation, romantic mystique, and intense program of indoctrination—produced an elitist mindset within the officer corps, a sense of separateness from and moral superiority to the mainstream of civilian society. Compared to the rowdy individualism, partisanship, and materialism that allegedly characterized the civilian world, the Military Academy . . . appeared an island of order, integrity, and devotion to duty."[18]

The professionalization of the antebellum officer corps was transformative not only for the army, but for other organizations. In New York, a similar observation could be made about West Pointers who came to the ASCE meetings at the Croton offices in Rotunda Park. William Sidell and George S. Greene, along with their civilian-trained engineering peers, sought to create an elitist mindset within the engineering profession. Not every canal laborer or artisan possessed the expertise to be called a "civil engineer." West Point officers in the antebellum army provided civilian-trained engineers an example of behavior and expectations for the profession to follow. Without West Point's graduates, the creation of the ASCE might very well have been delayed by a generation, or at least until after the Civil War, when other professions began to organize more formally through societies and associations.[19]

URBAN DEVELOPMENT

Another effect of the engineering curriculum at West Point was its well-timed influence on the development of nineteenth-century New York. Because Thayer and Mahan first provided cadets with a base of engineering knowledge in the 1820s and 1830s, that generation of graduates was able to complete its military service obligation and then, as civilians, secure engineering positions in the city. Consequently, Viele and Greene became proven engineering professionals in New York prior to the outbreak of the Civil War. Beyond the physical construction of Central Park and the Croton Aqueduct improvements, they also added to the ideas and scope of New York's future development, particularly in Viele's case. As an antebellum sanitary reformer, he heightened New York's awareness of other grand cities like Paris and London, where sewers helped stem the outbreak of cholera and other waterborne disease. By making public presentations and publishing articles and official reports, Viele became one of the voices advocating a way ahead for New York.

During and after the Civil War, Viele furthered French ideas in Victorian New York by urging that the city follow Louis Napoleon's and Baron Georges-Eugène Haussmann's example in Paris. With the support of his fellow members of the West Side Association, he recommended that New York expand with an eye toward grandeur, planning for tree-lined boulevards and modern apartment flats. Central Park and the Upper West Side with Riverside Drive were representative of this Francophile impulse. In creating grand spaces connected by elevated railways and lined by the tallest buildings technically possible, Viele and other city builders positioned New York to be unequaled by any other American city in the late nineteenth century.[20] Viele's Water Map provided a detailed, innovative understanding of the earth below Manhattan's blocks for builders to consult prior to erecting new skyscrapers. The West Side Association meetings at the Fifth Avenue Hotel were a forum for property owners, speculators, and boosters to discuss the planning and direction of Gotham's future boulevards, green spaces, and public transportation. They may not have had the absolute power that Haussmann wielded in Paris, but they all shared a vision for a greater New York, and that shared vision enabled the city to rival Paris and London as a world metropolis.[21]

In the United States, other urban centers were a distant second to New York City with respect to population, culture, and national influence. Boston may have been a "finished city"; Washington, DC, the political capital; and Chicago, the gateway to the West—but New York was the center of nineteenth-century American life. It was the hub of change in a modernizing society, or at least of any change driven by commerce, finance, and capitalism.[22] From the 1850s through the end of the century, Gotham's economic elite were the "largest and most diverse" in the United States, controlling over 71 percent of the city's wealth, higher than that in any other American city.[23] New York was the metropolis by which all other modern American cities were measured at the turn of the twentieth century.[24] The very feel of being in Gotham was like no other urban center in the United States.[25]

In the four decades after the Civil War, West Pointers, along with other engineers, contributed to the rise of a sublimely modern metropolis by applying their expertise to overcome geographic and environmental obstacles, and by managing New York's Public Works and Police Departments. Gotham's leaders came to depend upon the former military men to run the Streets Department, while Tammany and the Swallowtail Democrats looked to exploit their every move. As veteran Civil War generals, the West Point alumni approached the Police Board as they would have done an army staff. Instead of coordinating campaigns to defeat the Confederate Army, they applied their leadership skills to protect the populace from crime as well as police brutality. More than a decade before the start of Progressivism, Gotham expected them to reform any corruption and injustice associated with city government and services. As products of the nation's Military Academy, the alumni also had high expectations of themselves as they took on key positions in municipal government. But if they had hoped to improve the function and efficiency of the city's departments, they fell short of those expectations. The conservative tendencies inherent in the Military Academy and the army persisted in these officers even after they became civilians and pursued new goals while still accommodating New York's power brokers.

Their actions in postbellum New York were examples of what Morton Keller observed in his book *Affairs of State*. Keller argues that the Civil War did not change the character of America. Instead, it merely reinforced the "triumph of nationalism" and "the political power of the North" already ongoing in nineteenth-century American historical development.[26] However, "the Civil War generation had a profound sense of living through the events that changed their world." In many ways, the graduates of the Military Academy were an example of this "contradiction inherent in [the] legacy" of the Civil War.[27] They acted as though their world had been created anew but remained steadfast in their political views and affiliations. Their efforts led to preservation of the status quo in New York's social and political scene. The Academy alumni performed their duties within existing political systems and social relationships. As a consequence, these West Pointers marginally improved the political rights of the working class and newly arriving immigrants in the city. Changing the postwar world was not a priority for the generation of military men who had just gone through four years of traumatic national upheaval. If anything, they preferred to return to some sense of the familiar, and being Democrats in the city was a way to do so.

Beyond politics, though, West Pointers and their engineer brethren improved the quality of life for all New Yorkers, rich and poor, elites and working class. Croton water, a city sewer system, Central Park, Prospect Park, the Brooklyn Bridge, and public transportation all made Gotham an easier place to live and work. The advent of technology and its accompanying promise of a better future facilitated optimism and fueled the economic growth emanating from the city.[28] In New York, West Pointers were part of the engine of science and technology that created wealth in the second half of the nineteenth century in the United States.

George McClellan was the pivotal figure in determining the postbellum fates of West Pointers in New York. Postwar inclinations generally dictated that if an officer served with or under McClellan in the Army of the Potomac, then New York was more than likely the best place for civilian opportunities. McClellanites, as those loyal to the general were sometimes called, tended to be Democrats who came together in New York to take advantage of their experience, expertise, and affinities in a city that remained controlled by Democrats.[29] While the West Pointers never joined Tammany Hall, they remained loyal to the party, and were inclined to fall in with the Swallowtails. As veterans serving municipal needs, they could reaffirm or alter the public's view of events from the Civil War. Since McClellan remained a popular candidate after the Copperhead campaign of 1864, he and those who served under him found the political climate of postbellum New York to be receptive to their skills and professional qualities.

REDEMPTION

Coming to New York City was a means for former Union generals to find redemption for their failures and transgressions in the Civil War. They could reframe their wartime service as actors in the postbellum milieu of New York politics. Fitz John Porter's struggle, as one of the more public postwar redemptions, exemplified a sequence of events similar to those of McClellan, Slocum, and William F. Smith. The more they contributed to Gotham's development after the war, the more they could repudiate negative wartime legacies and recast their war records to garner greater respect, honor, and status. In the first two decades after the war, the city's embrace of McClellan's former commanders and supporters was profound. The popularity of John Logan's G.A.R. between 1865 and 1870, especially among Radical Republicans and radical sections of the nation, limited where Democratic veterans could find political and social acceptance.[30] Swallowtail Democrats like August Belmont and Abram Hewitt welcomed McClellan, William F. Smith, and Fitz John Porter as political allies who also possessed expertise that could be turned into political gain in New York. In Brooklyn, Slocum filled a unique niche as a veteran and notable Civil War leader whose renown made him a valuable asset to William C. Kingsley and the New York Bridge Company. Having a hero of Gettysburg on the board helped Kingsley and other board members maintain support for the Great Bridge.

In some cases, memorials commemorating wartime service were tangible signs of redemption for veterans associated with controversial events during the Civil War. For instance, Slocum suffered much public rebuke for his lack of Radical convictions in 1865 but ended up with two memorial statues after his death: one in Brooklyn's Prospect Park, built in 1905; and one at Gettysburg, dedicated in 1902.[31] It would be difficult to imagine Slocum having these memorials if he had remained in central New York and not come to Brooklyn as a newly converted Democrat in 1866. While Viele did not have a monument in the city, he did add to his legacy by remaining an active booster and advocate for the Upper West Side. He was among the celebrated veterans attending municipal celebrations over the last decade of the nineteenth century, and the Lafayette Post of the G.A.R. fondly remembered him after his death in 1902.[32] Fittingly for Viele, he had to make provisions for his own memorial and personally funded his pyramid tomb at West Point's cemetery. While the city has no memorial to Fitz John Porter, the court-martialed general did rehabilitate his reputation in New York and, with the bill reinstating him into the army, his supporters garnered enough support for a monument to be built in his hometown of Portsmouth, New Hampshire.[33]

But these memorials to the Democratic West Pointers in the city and elsewhere paled in comparison to other famous memorials dedicated to Union generals. Grant's Tomb, built on the Upper West Side in 1897, would be the most conspicuous symbol of the veterans' service. Since 1903, Gen. William T.

Sherman's statue has graced the entrance to Central Park, in the Grand Army of the Republic Plaza. New York erected statues in honor of Gouverneur Kemble Warren and Philip Sheridan in 1896 and 1936, respectively. The point was that, unlike the Democratic veteran generals, Grant, Sherman, Sheridan, and even Warren were memorialized for leading the U.S. Army to victory. Although all Union veterans could rightfully claim that they had supported the North's efforts, popular opinion viewed Grant and Sherman as the real saviors of the Union since they had led those last campaigns of 1865.[34] Postwar Americans recognized Grant and Sherman as national heroes and thus their memorials in the city served more as national cenotaphs than as municipal recognitions of their actions.[35] Of the Democratic generals, perhaps Slocum had the greatest success at revising his legacy, as Brooklyn's leaders placed his statue in the Grand Army of the Republic Plaza in Prospect Park.

NEW YORK AND AMERICA ON THE EVE
OF THE TWENTIETH CENTURY

Regardless of whom the monuments of bronze, marble, and stone commemorated, they stood like cairns guiding New York and the nation to remembrance of the sacrifice and heroics of the Civil War generation led by the graduates of the Military Academy. Understandably, the Civil War veterans wanted assurance that their legacy would last well into the next century. Their monuments also affected how Americans conceived of themselves at the turn of the twentieth century. By defeating the Confederacy, Grant, Sherman, Slocum, and other former Union generals exorcised the Old South and its encumbrances from American consciousness. Later, when they became civilian engineers, commissioners, politicians, and heroes in New York, they perpetuated a desirable interpretation of the Civil War—one in which the United States removed the stains of slavery, not only from the South, but also from the national economy and commercial wealth that emanated from New York. The war eliminated undesirable scars that had come from its antebellum associations with the "cotton kingdom," and the North allowed the postwar nation to pursue a different destiny. Statues of Civil War generals in New York reflected the impulse to remember these leaders, and they also served as permanent reminders that the North had corrected the nation's errant course by smashing the Confederacy and forcing an end to slavery.

Frederick Jackson Turner's frontier thesis encapsulated the moment by emphasizing the drive for westward expansion. Despite his focus on the West, he still recognized that the East had a central role, even if it was to "check and guide" that territorial growth. Turner described his interpretation of the relationship between the East and the West after the Civil War when he wrote:

> Magnitude of social achievement is the watchword of the democracy since the Civil War. From petty towns built in the marshes, cities arose whose greatness and industrial power are the wonder of our time. The conditions were ideal for the production of captains of industry. The old democratic admiration for the self-made man, its old deference to the rights of competitive individual development, together with the stupendous natural resources that opened to the conquest of the keenest and the strongest, gave such conditions of mobility as enabled the development of the large corporate industries which in our own decade have marked the West.[36]

The geographic center of the United States may have been the Midwest in 1893, but New York was still the core of America socially, culturally, economically, and politically.[37] Gotham was the engine of finance and commerce, driving the nation west after the Civil War.

In this context, Military Academy alumni maintained the New York that was the core of progress and change. They applied their experience and expertise to the challenges of an ever expanding and evolving metropolis. Whether as engineers, commissioners, or both, the West Pointers reflected in their actions a firm foundation in the engineering curriculum advocated by Dennis Hart Mahan. Where Viele saw disease-inducing sanitation conditions, he used topographical drawings to show how to mitigate the problem. Greene, Newton, Lovell, and G. W. Smith used their engineering know-how to bring water to Manhattan and to clear New York's waterways. To run the Police Board, Fitz John Porter and William F. Smith used organizational leadership skills first learned on the plain at West Point. Throughout the nineteenth century, each graduate attempted through engineering and science to emplace order over an increasingly disordered metropolis.

A version of order did come, but only after the era of West Pointers in the city. Theodore Roosevelt, George Waring, and the Progressives appropriated much of the Military Academy alumni's example in pursuing Progressive reform.[38] Elements of natural and social sciences, military action, and ideals of duty and commitment to country became part of the Progressive vernacular. Progressives singing "Onward, Christian Soldiers!" at the national convention of 1912 was just one such example.[39] Indeed, "the wartime frenzy of idealism and self-sacrifice marked the apotheosis as well as the liquidation of the Progressive spirit" during World War I.[40] Instead of looking to previous practices and working within existing social and political frameworks as the West Pointers had done in their peacetime pursuits, the Progressives considered more comprehensive methods to eradicate the problems they saw created by modernization and industrialization. Yet those problems were not necessarily new, and in Victorian New York, West Pointers were effective at mitigating the obstacles and hindrances to the city's development from 1833 through the 1880s.

As early as 1865, New York's future as a modern city of the world was predictable. Egbert L. Viele reasoned, "New palaces of trade and industry are rising up on every hand, and so it will go on; capital will continue to seek here an investment, and labor its reward, and in a few years will find a city rivaling in population, opulence, and splendor, any city of ancient or modern times."[41]

As George Templeton Strong had noted in 1871, Viele was right to predict that his New York would be a magnificent new city. However, it would take thirty years for the transformation to be complete. The change that he saw around him was the work of West Point professionals collaborating with their fellow civilian engineers, boosters, and politicians in Gotham. Given the progress noticeable at the end of the 1860s, visions of the future were easily conceived. Civil engineering had created great anticipation among New Yorkers for a new metropolis to rise. Greene, Viele, Newton, Slocum, and others were part of a much longer and gradual urban transformation that had started in the antebellum era and was realized at the turn of the twentieth century. In Gotham, the Long Gray Line helped lay the foundation for the ideal American city, the engineering profession, and the veneration of science and technology in the century to come. Ultimately, that foundation proved to be more important than their romanticized feats in the American West. In New York, perhaps, West Point's engineers made their greatest contribution to the nation.

ACKNOWLEDGMENTS

As with most callings in life, it is no coincidence that my interests and research turned to New York City. While I was born at one end of the Erie Canal in Buffalo, my mother's parents began their life together at the other end of the old canal route, in New York City. As my grandmother related the story to me years ago, she and my grandfather had taken a December trip to the city, and he proposed to her during a carriage ride in Central Park. Some sixty years later, it was Central Park that first sparked my inquiry into how men from the U.S. Military Academy could have played a role in creating a special swath of nature in the most urban place on earth. Over the years, my project evolved and changed many times, but invariably, the city, the park, and the men who shaped both have remained the touchstone of this project. While I am indebted to many people for their support, guidance, and understanding along the way, I suppose my first expression of gratitude should go to Central Park and its enduring inspiration to lovers and scholars alike. Without it, neither I nor this work would exist.

My research into the relationship between the men of West Point and New York City began while I was an American History instructor at the U.S. Military Academy. One of my additional duties at the Academy was to provide tours of the West Point Cemetery and the various figures interred there. At the pyramid tomb of Egbert L. Viele, I had to explain that he was a graduate from the class of 1847, served in the Mexican-American War, was the first designer of Central Park, and was a Union general in the Civil War. It was the phrase "the first designer of Central Park" that left me concerned that I might be embellishing Viele's legacy, especially given the rightful place of Frederick Law Olmsted and Calvert Vaux in the history of the park. So in an effort not to be embarrassed while leading tours, I began my research. Since the fall of 2001, the people who have supported me in this project are many.

At LSU Press, I am grateful that Mike Parrish and Rand Dotson took on this manuscript. During the editing process, Mike fed me a steady stream of relevant books, articles, and ideas to broaden my vision and put this topic into context. This book would not have been possible without his patient guidance, wealth of publishing experience, and vast understanding of the history of the United States in the nineteenth century. Rand provided me ample space and the necessary amount of nudging to get the manuscript "right" for publication.

I am indebted to Colonel (retired) Gary Tocchet, and Brigadier General (retired) Lance Betros for supporting my initial research and a paper I delivered at the West Point Bicentennial History Conference in the spring of 2002. Chase Viele, a great-nephew of Egbert Viele, pointed me to many valuable sources through correspondence and phone conversations. Alan Aimone of Special Collections at the USMA Cadet Library provided invaluable assistance with West Point sources, especially in the early stages of research.

For nearly five years, this project stalled as duty called me to Korea, Afghanistan, and Iraq. Throughout that time, Patti Bohrer in the Syracuse University History Department kept me "alive" academically, ensuring that I maintained good status with the university as I circled the world for the army. Of course, Professor Margaret Susan (Peggy) Thompson was my constant patron, waiting for me to return to a post where I could write, sending notes of encouragement while I was deployed, and keeping me on track when I resumed working on this project in 2008. This study would never have been completed without her sage advice and understanding friendship.

I am also indebted to the many dedicated faculty in the History Department at Syracuse. David Bennett and Scott Strickland were instrumental in forming my approach to the study of history. David showed me how powerful a well-researched narrative could be. Scott first opened my eyes to the world of nineteenth-century America. They made this journey of discovery enriching and rewarding.

The administration and fellow faculty members at the U.S. Naval War College in Newport enabled me to pick up where I left off and to mature as a historian. John Maurer, then chair of the Strategy and Policy Department, gave me the time and the travel funds to finish my research. Kevin McCranie read more than his share of draft chapters and gave me "brutal but honest" feedback that always improved whatever I sent him. Michael Vlahos routinely expanded my vision for the project, and his infectious motivation kept me going when I thought I could not. Paul Krajeski, my colleague and friend, provided timely encouragement and was usually the first audience for my chapter ideas and research findings. Douglas V. Smith and Stanley Carpenter actively cheered me on and urged me to keep going. Jeff Shaw provided valuable comments and supported my efforts throughout.

I owe the War Gaming Department chair, Dave DellaVolpe, a deep debt of gratitude for taking me on board and allowing me the latitude to complete this book. My colleagues in the War Gaming Department have been supportive, providing enthusiastic encouragement to see this book to the end. Hank Brightman and Doug Ducharme were a constant source of perspective and reassurance in the final stages of the project. The men and women who study, research, and teach at Newport represent the very best of America's navy professionals. I am proud to work with them all.

Last but not least at the Naval War College, Robin Lima of the Henry E. Eccles Library found every source I requested through interlibrary loans. Her dedication and perseverance saved me multiple trips to distant archives for primary sources. There are many more colleagues and staff at the Naval War College whom I could mention, but there is not enough space here.

During my time in New York City, I stayed on Long Island with my mother-in-law and her husband, and later with a former colleague, Patrick Smith. As I played "New York commuter" during one of my weeks of research, I planned to take the Long Island Railroad into Penn Station every day. On the first day of my commute, a hundred-year-old switch burned up at Jamaica Station, delaying all Long Island trains for three hours and permanently ending all express trains for the remainder of my visit. To quote one of my hosts, "You picked a hell of a week to be in New York." Thus, my commute of two hours and twenty minutes turned into a daily three-hour ordeal. After that week of research and the full experience of being a New York commuter, I vowed that I would seek only to write about New York, but never attempt to make a living there.

Many friends and associates reviewed draft chapters along the way, but I want to give special thanks to Ted Crackel for his comments and advice, and to Frank Grzyb for his editorial assistance.

Finally, I want to thank my wife Peg and our children, Charlotte and Ben. Peg helped me balance the urgency to get this book done with the demands, and desires, of being in a wonderful family with young children. I alternated Saturday mornings on soccer fields and baseball diamonds with weekends in the basement. Bedtime readings of *Curious George* and *Clifford the Big Red Dog* were often followed by hours of examining Mahan's textbooks and nineteenth-century penmanship. When I had my moments of doubt and frustration, Peg pushed me to get this done. She has been steadfast with her love and support. Like the history in these pages, our marriage and family began at West Point. I dedicate this book to her and the kids.

NOTES

INTRODUCTION:
NINETEENTH-CENTURY NEW YORK AND WEST POINT

1. Just two months after Strong's diary entry, Mahan took his life on 16 September 1871, by jumping into the paddle wheel of a Hudson River steamship. See Henry L. Abbott, "Memoir of Dennis Hart Mahan: Read before the National Academy, November 7, 1878," *Memoirs of the National Academy of Sciences* (Washington, DC: Government Printing Office, 1886), 2:35.

2. While there are several pronunciations for the name "Viele," Egbert more than likely pronounced it "veal" and was sometimes called "General Weal." See Egbert L. Viele, "Lincoln as Storyteller," in William Hayes Ward, *Abraham Lincoln: Tributes from His Associates, Reminiscences of Soldiers* (New York: Thomas Y. Crowell, 1895), 120.

3. George J. Lankevich, *New York City: A Short History* (New York: New York University Press, 2002), 67, 126.

4. *Scientific American* 8.6 (23 October 1852), 45.

5. Herbert Croly, "New York as the American Metropolis," *Architectural Record* 13 (1903), 194.

6. Croly, 195; also see Eugene P. Moehring, *Public Works and the Patterns of Urban Real Estate Growth in Manhattan, 1835–1894* (New York: Arno, 1981), 1–15.

7. Edward L. Glaeser, "Urban Colossus: Why New York Is America's Largest City," *Federal Reserve Bank of New York Economic Policy Review* 11.2 (2005): 7–24.

8. Edwin G. Burrows and Mike Wallace, *Gotham: A History of New York City to 1898* (New York: Oxford University Press, 1999). *Gotham* is the best single-volume monograph on the history of New York City. In an informative and well-documented narrative, Burrows and Wallace detail the history of the city from Dutch settlement through the incorporation of the five boroughs into Greater New York in 1898. Other useful accounts of New York in the nineteenth century include Sean Wilentz, *Chants Democratic: New York City and the Rise of the American Working Class, 1788–1850* (New York: Oxford University Press, 1986), Alexander Callow, *The Tweed Ring* (New York: Oxford University Press, 1969), Roy Rosenzweig and Elizabeth Blackmar, *The Park and the People: A History of Central Park* (Ithaca, NY: Cornell University Press, 1992), David McCullough, *The Great Bridge: The Epic Story of the Building of the Brooklyn Bridge* (New York: Simon and Schuster, 1982), and David Scobey, *Empire City: The Making and Meaning of the New York City Landscape* (Philadelphia: Temple University Press, 2002). Kenneth Jackson, ed., *The Encyclopedia of New York*

City (New Haven, CT: Yale University Press, 1995), has become an indispensable standard reference for all things New York, past and present.

9. James R. Endler, *Other Leaders, Other Heroes: West Point's Legacy to America beyond the Field of Battle* (Westport, CT: Praeger, 1998), 63–73.

10. See Todd Shallat, *Structures in the Stream: Water, Science, and the Rise of the U.S. Army Corps of Engineers* (Austin: University of Texas Press, 1994), 78–116.

11. For a sample of West Point histories, see Endler; R. Ernest Dupuy, *Where They Have Trod: The West Point Tradition in American Life* (New York: Frederick A. Stokes, 1940); and Theodore J. Crackel, *West Point: A Bicentennial History* (Lawrence: University Press of Kansas, 2002). For New York City histories, see note 7.

12. William B. Skelton, *An American Profession of Arms: The Army Officer Corps, 1784–1861* (Lawrence: University Press of Kansas, 1992), 172, 400n17.

13. Carl Smith, *City Water, City Life: Water and the Infrastructure of Ideas in Urbanizing Philadelphia, Boston, and Chicago* (Chicago: University of Chicago Press, 2013), 2.

14. See Catherine McNeur, *Taming Manhattan: Environmental Battles in the Antebellum City* (Cambridge, MA: Harvard University Press, 2014), 4–5; and Jackson Lears, *Rebirth of a Nation: The Making of Modern America, 1877–1920* (New York: Harper Perennial, 2010), 1–6.

15. George Templeton Strong, *The Diary of George Templeton Strong,* vol. 2, *The Turbulent Fifties, 1850–1859,* ed. Allan Nevins and Milton Halsey Thomas (New York: Macmillan, 1952), 55–56.

16. Stephen E. Ambrose, *Duty, Honor, Country: A History of West Point* (Baltimore: Johns Hopkins University Press, 1999), 97. This earlier work by Ambrose is well documented and appears to avoid the errors that plague the historian's later efforts. Also see Mark Aldrich, "Earnings of American Civil Engineers, 1820–1859," *Journal of Economic History* 31.2 (June 1971): 407–19.

17. George S. Pappas, *To the Point: The United States Military Academy, 1802–1902* (Westport, CT: Praeger, 1993), 275; James D. Dilts, *The Great Road: The Building of the Baltimore and Ohio, The Nation's First Railroad, 1828–1853* (Stanford, CA: Stanford University Press, 1993), 63.

18. Endler, 67; USMA Association of Graduates, *Register of Graduates and Former Cadets of the United States Military Academy, Class of 1900 Centennial Edition* (West Point, NY, 2000), Cullum no. 327.

19. Rosenzweig and Blackmar, 100–102.

20. USMA Association of Graduates, *Annual Reunion of the Association of Graduates of the United States Military Academy,* 9 June 1902, 143; *Biographical Directory of the United States Congress, 1774–Present,* accessed 19 October 2015 at http://bioguide.congress.gov/.

21. *Biographical Directory of the United States Congress, 1774–Present;* McCullough, *The Great Bridge.* McCullough includes Slocum throughout his narrative of the Roeblings and the Brooklyn Bridge.

22. See Scobey; Burrows and Wallace; Edward K. Spann, *The New Metropolis: New York City, 1840–1857* (New York: Columbia University Press, 1981); David C. Hammack, *Power and Society: Greater New York at the Turn of the Century* (New York: Russell Sage Foundation, 1982); and Rosenzweig and Blackmar.

23. Scobey, 13–14. In correspondence of 20 March 2002, Scobey admitted that he "should have" dealt more with Viele, who "has been unjustly neglected."

24. Ibid., 7, 14.

25. Burrows and Wallace, 950.

26. Peter Salwen, *Upper West Side Story: A History and Guide* (New York: Abbeville, 1989), 60, 72.

27. Fitz John Porter to William H. Wickham, 30 April 1875, in Samuel J. Tilden Papers, New York Public Library MSS, Box 20.54.

28. Crackel, 1.

29. See ibid.; James L. Morrison Jr., *"The Best School": West Point, 1833–1866* (Kent, OH: Kent State University Press, 1998); and Pappas.

30. Barton C. Hacker, *American Military Technology: The Life Story of a Technology* (Westport, CT: Greenwood, 2006), 7–10.

31. Hugo A. Meier, "Technology and Democracy, 1800–1860," *Mississippi Valley Historical Review* 44.2 (1957): 624.

32. Shallat, 2–9; Thomas Kessner, *Capital City: New York City and the Men behind America's Rise to Economic Dominance, 1860–1900* (New York: Simon and Schuster, 2003), 53–54.

33. Forest G. Hill, *Roads, Rails, & Waterways: The Army Engineers and Early Transportation* (Norman: University of Oklahoma Press, 1957), 3, 204–26.

34. Strong, *The Diary of George Templeton Strong*, vol. 1, *Young Man in New York, 1835–1849*, ed. Nevins and Thomas, 356–63.

35. Pappas, 242–43.

36. Marcus Cunliffe, *Soldiers and Civilians: The Martial Spirit in America, 1775–1865* (London: Erye & Spottiswoode, 1968), 101. Also see Samuel J. Watson, "How the Army Became Accepted: West Point Socialization, Military Accountability, and the Nation-State during the Jacksonian Era," *American Nineteenth Century History* 7 (June 2006): 221.

37. Jean H. Baker, *Affairs of Party: The Political Culture of Northern Democrats in the Mid-Nineteenth Century* (New York: Fordham University Press, 1998), 154.

I. WEST POINT INFLUENCE IN VICTORIAN GOTHAM

1. As will be discussed in chapter 4, civil-trained engineers came from a master-apprentice system used extensively by engineers building the Erie Canal system.

2. Edwin G. Burrows and Mike Wallace, *Gotham: A History of New York City to 1898* (New York: Oxford University Press, 1999), 966–69.

3. David W. Palmer, *The Forgotten Hero of Gettysburg: A Biography of General George Sears Greene* (Philadelphia: Xlibris, 2005), 48; Edward Wegmann, *The Water-Supply of the City of New York, 1658–1895* (New York: John Wiley & Sons, 1896), 112–13.

4. Roy Rosenzweig and Elizabeth Blackmar, *The Park and the People: A History of Central Park* (Ithaca, NY: Cornell University Press, 1992), 100–101. In 1875, after the state legislature returned control of the New York Police Department board to the city, the mayor appointed William F. Smith as a police commissioner.

5. James R. Endler, *Other Leaders, Other Heroes: West Point's Legacy to America beyond the Field of Battle* (Westport, CT: Praeger, 1998), 63–73.

6. See William B. Skelton, *An American Profession of Arms: The Army Officer Corps, 1784–1861* (Lawrence: University Press of Kansas, 1992); Marcus Cunliffe, *Soldiers and Civilians: The Martial*

Spirit in America, 1775–1865 (London: Erye & Spottiswoode, 1968); and Samuel P. Huntington, *The Soldier and the State: The Theory and Politics of Civil-Military Relations* (Cambridge, MA: Belknap Press of Harvard University Press, 1957, 1985).

7. Skelton, *An American Profession of Arms,* xiii.

8. Ibid., xiv; Huntington, 161.

9. Huntington, 8, 9, 194, 211–21.

10. Burton J. Bledstein, *The Culture of Professionalism: The Middle Class and the Development of Higher Education in America* (New York: Norton, 1978), 193.

11. William B. Skelton, "Samuel P. Huntington and the Roots of the American Military Tradition," *Journal of Military History* 60.2 (April 1996): 334.

12. Huntington, 161.

13. Frank J. Walton, "The West Point Centennial: A Time for Healing," in *West Point: Two Centuries and Beyond,* ed. Lance Betros (Abilene, TX: McWhiney Foundation Press, 2004), 209–14; J. T. Seidule, "'Treason Is Treason': Civil War Memory at West Point, 1861–1902," *Journal of Military History* 76.2 (2012): 427–545.

14. Jacob Kobrick, "No Army Inspired: The Failure of Nationalism at Antebellum West Point," *Concept: An Interdisciplinary Journal of Graduate Studies* (Villanova University, 2004): 4, 17.

15. Many excellent studies chronicle the internal improvements of the United States that were supported by graduates of the Military Academy between 1820 and 1900. For example, see James D. Dilts, *The Great Road: The Building of the Baltimore and Ohio, the Nation's First Railroad, 1828–1853* (Stanford, CA: Stanford University Press, 1993); Forest G. Hill, *Roads, Rails, & Waterways: The Army Engineers and Early Transportation* (Norman: University of Oklahoma Press, 1957); Robert G. Angevine, *The Railroad and the State: War, Politics, and Technology in Nineteenth-Century America* (Stanford, CA: Stanford University Press, 2004); Stephen Ambrose, *Nothing like It in the World: The Men Who Built the Transcontinental Railroad, 1863–1869* (New York: Touchstone, 2000); and Aubrey Parkman, *Army Engineers in New England: The Military and Civil Work of the Corps of Engineers in New England, 1775–1975* (Waltham, MA: U.S. Army Corps of Engineers, 1978).

16. Paul E. Cohen and Robert T. Augustyn, *Manhattan in Maps, 1527–1995* (New York: Rizzoli, 1997), 101–103; Kenneth T. Jackson, *Crabgrass Frontier: The Suburbanization of the United States* (New York: Oxford University Press, 1985), 74–75.

17. Leonard P. Curry, *The Corporate City: The American City as a Political Entity, 1800–1850* (Westport, CT: Greenwood, 1997), 184.

18. Alexander Hamilton Jr. to his Grandfather Robert Morris, 27 April 1843, in Misc. MSS, Alexander Hamilton Jr., Letters, 1823–1882, New-York Historical Society, Folder 1.

19. Glenn Collins, "Birth of Central Park Holds Parallels with Ground Zero," *New York Times,* 15 May 2003, B1.

20. Joanne Abel Goldman, *Building New York's Sewers: Developing Mechanisms of Urban Management* (West Lafayette, IN: Purdue University Press, 1997), 152–66.

21. David McCullough, *The Great Bridge: The Epic Story of the Building of the Brooklyn Bridge* (New York: Simon and Schuster, 1982), 550–52.

22. Mona Domosh, *Invented Cities: The Creation of Landscape in Nineteenth-Century New York and Boston* (New Haven, CT: Yale University Press, 1996), 8–9.

23. James Fenimore Cooper, "The Towns of Manhattan" (unpublished, 1851), excerpted in *Empire City: New York through the Centuries,* ed. Kenneth T. Jackson and David S. Dunbar (New York: Columbia University Press, 2002), 150.

24. David Scobey, *Empire City: The Making and Meaning of the New York City Landscape* (Philadelphia: Temple University Press, 2002), 5–7.

25. Ibid., 8. See Daniel Schaffer, ed., *Two Centuries of American Planning* (Baltimore: Johns Hopkins University Press, 1988).

26. Scobey, 8.

27. Egbert L. Viele, "Report on the Civic Cleanliness, and the Economical Disposition of the Refuse of Cities" (New York: Edmund Jones, 1860), 7, 8.

28. *Real Estate Record and Builders' Guide* 28 (26 November 1881).

29. To get a sense of Viele's hubris and self-importance, see his essay "Lincoln as Story Teller," in *Abraham Lincoln: Tributes from His Associates, Reminiscences of Soldiers, Statesmen, and Citizens,* ed. William Hayes Ward (New York: Thomas Y. Crowell, 1895), 116–24. In that essay, Viele claims, "From that time until Mr. Lincoln's death I enjoyed the very closest intimacy with him." Surveying the vast literature published on the sixteenth president, one would be hard pressed to corroborate Viele's claim.

30. Stephen W. Sears, *George B. McClellan: The Young Napoleon* (New York: Ticknor and Fields, 1988), 387–88, 393.

31. Sven Beckert, *The Monied Metropolis: New York City and the Consolidation of the American Bourgeoisie, 1850–1896* (New York: Cambridge University Press, 2001), 3.

32. "A Monument to Slocum," *Brooklyn Eagle,* 30 April 30 1894, 9.

33. Daniel M. Bluestone, "From Promenade to Park: The Gregarious Origins of Brooklyn's Park Movement," *American Quarterly* 39.4 (Winter 1987): 542.

34. Rosenzweig and Blackmar, 309.

35. USMA Association of Graduates, *Annual Reunion of the Association of Graduates of the United States Military Academy,* 10 June 1895, 110.

36. See David Quigley, *Second Founding: New York City, Reconstruction, and the Making of American Democracy* (New York: Hill and Wang, 2004), xiv–xv; and Anthony Gronowicz, *Race and Class Politics in New York City before the Civil War* (Boston: Northeastern University Press, 1998), x–xviii.

37. Burrows and Wallace, 1220–26.

2. AN AMERICAN ECOLE POLYTECHNIQUE

1. George Ticknor to his wife, Mrs. Ticknor, 10 June 1826, in George Ticknor, *Life, Letters, and Journals of George Ticknor* (Boston: James R. Osgood, 1877), 1:375.

2. Board of Visitors, *Annual Report of the Board of Visitors to the United States Military Academy* (June 1824), 100.

3. Thomas E. Griess, "Dennis Hart Mahan: West Point Professor and Advocate of Military Professionalism, 1830–1871" (PhD diss., Duke University, 1968), 111–12.

4. See ibid., 206–207; Stephen E. Ambrose, *Duty, Honor, Country: A History of West Point*

(Baltimore: Johns Hopkins University Press, 1999), 99–100; and Theodore J. Crackel, *West Point: A Bicentennial History* (Lawrence: University Press of Kansas, 2002), 125.

5. Griess, 206–207.

6. Alexa James, "West Point Aids Afghan Counterpart; New Academy Has 1st Grads," *Times Herald-Record,* 2 February 2009, accessed 6 November 2015 at http://www.recordonline.com/apps /pbcs.dll/article?AID=/20090202/NEWS/902020319/-1/ . . .

7. "The 2000 Campaign: Transcript of Debate between Vice President Gore and Governor Bush," *New York Times,* 4 October 2000. President George W. Bush, prior to taking office, campaigned on a policy that the United States would no longer "nation-build" with its military.

8. Richard M. Johnson, "Report on the Military Academy at West Point, by the Committee on Military Affairs, Submitted by Its Chairman, Col. R. M. Johnson," 17 May 1834, in *American Quarterly Review* 16 (Sept.–Dec. 1834): 358–75.

9. Francis Wayland, *Report to the Corporation of Brown University, on Changes in the System of Collegiate Education* (Providence, RI: George H. Whitney, 1850), 18. In his address, read 28 March 1850, Brown University president Francis Wayland remarked, "We presume the single academy at West Point, graduating annually a smaller number than many of our colleges, has done more towards the construction of railroads than all our one hundred and twenty colleges united." Many studies assert the role of West Point graduates in building the internal improvements of the United States in the nineteenth century. Daniel H. Calhoun's *The American Civil Engineer: Origins and Conflict* (Cambridge, MA: Harvard University Press, 1960) is one of the most cited works. Robert G. Angevine's *The Railroad and the State: War, Politics, and Technology in Nineteenth-Century America* (Stanford, CA: Stanford University Press, 2004) is a more recent study focusing on the construction of railroads. The general consensus of the historiography supports Wayland's remarks a century and a half ago.

10. Todd A. Shallat, *Structures in the Stream: Water, Science, and the Rise of the U.S. Army Corps of Engineers* (Austin: University of Texas Press, 1994), 114.

11. James L. Morrison Jr., *"The Best School": West Point, 1833–1866* (Kent, OH: Kent State University Press, 1998), ix–x.

12. Crackel, 1.

13. Major Jonathan Williams to Major Decius Wadsworth, 13 August 1802, Jonathan Williams Papers, quoted in Peter Michael Molloy, "Technical Education and the Young Republic: West Point as America's Ecole Polytechnique, 1802–1833" (PhD diss., Brown University, 1975), 241–42.

14. Ambrose, *Duty, Honor, Country,* 88–89.

15. Brian M. Linn, *The Echo of Battle: The Army's Way of War* (Cambridge, MA: Harvard University Press, 2007), 14–15.

16. Griess, 217–19.

17. Daniel Walker Howe, *What Hath God Wrought: The Transformation of America, 1815–1848* (New York: Oxford University Press, 2007), 749.

18. Wayne Wei-siang Hsieh, *West Pointers and the Civil War: The Old Army in War and Peace* (Chapel Hill: University of North Carolina Press, 2009), 91; Jacob Kobrick, "No Army Inspired: The Failure of Nationalism at Antebellum West Point," *Concept: An Interdisciplinary Journal of Graduate Studies* (Villanova University, 2004): 4; Morrison, *"The Best School,"* 101.

19. See Hsieh, 90–92; Crackel, 132–45; and Morrison, *"The Best School,"* 99–110.

20. James Morrison, "The Struggle between Sectionalism and Nationalism," *Civil War History* 19 (June 1973): 143; Clay Mountcastle, *Punitive War: Confederate Guerrillas and Union Reprisals* (Lawrence: University Press of Kansas, 2009), 19; Kobrick, 4, 8.

21. William B. Skelton, *An American Profession of Arms: The Army Officer Corps, 1784–1861* (Lawrence: University Press of Kansas, 1992), xiv.

22. Marcus Cunliffe, *Soldiers and Civilians: The Martial Spirit in America, 1775–1865* (London: Erye & Spottiswoode, 1968), 352, 358, 387–88. Cunliffe's study was for officers of all branches, not just those in the Corps of Engineers.

23. Samuel Watson, "Historiographical Essay: Continuity in Civil-Military Relations and Expertise: The U.S. Army before the Civil War," *Journal of Military History* 75.1 (January 2011): 231, 238, 248.

24. Elizabeth Brown Pryor, *Reading the Man: A Portrait of Robert E. Lee through His Private Letters* (New York: Penguin, 2008), 67; Peter Onuf, "Introduction," and Samuel Watson, "Developing '"Republican Machines"': West Point and the Struggle to Render the Officer Corps Safe for America, 1802–33," in *Thomas Jefferson's Academy: Founding West Point*, ed. Robert M. S. McDonald (Charlottesville: University of Virginia Press, 2004), 1–20, 154–74.

25. Burton J. Bledstein, *The Culture of Professionalism: The Middle Class and the Development of Higher Education in America* (New York: Norton, 1978), 195.

26. Dennis Hart Mahan, *A Treatise on Field Fortification, Containing Instructions on the Methods of Laying Out, Constructing, Defending, and Attacking Intrenchments, with the General Outlines Also of the Arrangement, the Attack and Defence of Permanent Fortifications* (New York: John Wiley, 1836, 1862), 46.

27. Morrison, *"The Best School,"* 153; Major Jonathan Williams, quoted in Peter Michael Molloy, "Technical Education and the Young Republic: West Point as America's Ecole Polytechnique, 1802–1833" (PhD diss., Brown University, 1975), 241–42.

28. George S. Pappas, *To the Point: The United States Military Academy, 1802–1902* (Westport, CT: Praeger, 1993), 32.

29. David F. Noble, *The Religion of Technology: The Divinity of Man and the Spirit of Invention* (New York: Alfred A. Knopf, 1997), 81–86.

30. Alan I. Marcus and Howard P. Segal, *Technology in America: A Brief History* (New York: Harcourt Brace Jovanovich, 1989), 88–89.

31. A. J. Angulo, "The Polytechnic Comes to America: How French Approaches to Science Instruction Influenced Mid-Nineteenth Century American Higher Education," *History of Science* 50, part 3 (September 2012): 318–19.

32. Sylvanus Thayer to Secretary of War Barbour, about 1826, quoted in Molloy, 432.

33. Morrison, *"The Best School,"* 111; Angulo, 329.

34. Paul Keith Nienkamp, "A Culture of Technical Knowledge: Professionalizing Science and Engineering Education in Late-Nineteenth Century America" (PhD diss., Iowa State University, 2008), 16. Also see Daniel Calhoun's *The American Civil Engineer*, 24–54. Calhoun notes that prior to 1837, the preponderance of civil engineers "consisted mainly of men with no school training as engineers . . . who had worked up within the engineer corps of internal improvement." However,

"the 'best' engineers in the United States of that period" came from West Point and schools like Rensselaer and Norwich. The other producer of engineers in the early national period was the New York canal system between 1814 and 1826 (53–54).

35. Terry S. Reynolds, "The Education of Engineers in America before the Morrill Act of 1862," *History of Education Quarterly* 32.4 (Winter 1992): 459–82.

36. Forest G. Hill, *Roads, Rails, & Waterways: The Army Engineers and Early Transportation* (Norman: University of Oklahoma Press, 1957), 208.

37. Larry Manning, "The Contribution of Sylvanus Thayer and the United States Military Academy to Engineering Programs in the United States" (PhD diss., Texas A&M University, 2003), 246–47.

38. John Rae and Rudi Volti, *The Engineer in History* (New York: Peter Lang, 2001), 182. Although not a land grant university, Yale University was the first to award a PhD in engineering in 1861.

39. Manning, 246–47.

40. Molloy, 411.

41. Reynolds, 467–68.

42. See Robert McDonald, ed., *Thomas Jefferson's Academy: Founding West Point* (Charlottesville: University of Virginia Press, 2004).

43. Ambrose, *Duty, Honor, Country,* 18.

3. THE ACADEMY OF THAYER AND MAHAN

1. James L. Morrison Jr., *"The Best School": West Point, 1833–1866* (Kent, OH: Kent State University Press, 1998), 15; Larry Manning, "The Contribution of Sylvanus Thayer and the United States Military Academy to Engineering Programs in the United States" (PhD diss., Texas A&M University, 2003), 131–32.

2. Sylvanus Thayer to Joseph Swift, 10 October 1815, in *The West Point Thayer Papers, 1802–1872,* ed. Cindy Adams (Association of Graduates, 1965), section 2.

3. Peter Michael Molloy, "Technical Education and the Young Republic: West Point as America's Ecole Polytechnique, 1802–1833" (PhD diss., Brown University, 1975), 367.

4. John Rae and Rudi Volti, *The Engineer in History* (New York: Peter Lang, 2001), 178.

5. The Alden Partridge affair is well known to scholars of West Point. Selected as the first superintendent of the Military Academy who was *not* also the chief of engineers in 1815, Partridge led the Academy with little support from the faculty. During his tenure, he sought to make the curriculum more focused on military training and less on academics. There was no standardized academic year or program of study. Partridge practiced favoritism among the Corps of Cadets, which resulted in numerous accusations of scandal and fraud. Eventually, Partridge could not overcome the mounting allegations against him, and President Monroe ordered Thayer to relieve Partridge. When Thayer arrived at West Point, Partridge refused to leave until a court-martial ordered his removal. The best account of the story is in Stephen E. Ambrose, *Duty, Honor, Country: A History of West Point* (Baltimore: Johns Hopkins University Press, 1999), 38–61. Partridge was a graduate of the class of 1806.

6. Theodore J. Crackel, *West Point: A Bicentennial History* (Lawrence: University Press of Kansas, 2002), 76–80. Crackel details the whole affair and controversy as the Congress and President Madison moved the Academy out from under the purview of the Army Corps of Engineers.

7. Robert F. Hunter and Edward L. Dooley Jr., *Claudius Crozet: French Engineer in America, 1790–1864* (Charlottesville: University Press of Virginia, 1989), 10–11, 14–17.

8. Molloy, 370–71.

9. Hunter and Dooley, 20–30.

10. Morrison, *"The Best School,"* 23.

11. William P. Leeman, *The Long Road to Annapolis: The Founding of the Naval Academy and the Emerging American Republic* (Chapel Hill: University of North Carolina Press, 2010), 75–76.

12. Crackel, 95–96; *Annual Report of the Board of Visitors to the United States Military Academy* (1824), 16.

13. Crackel, 1, 84, 96.

14. Rufus King to Christopher Gore, 22 June 1821, in *Life and Correspondence of Rufus King*, ed. Charles R. King (New York: Putnam and Sons, 1900), 6:393–94.

15. *Debates and Proceedings in the Congress of the United States, 1834–1856* (Washington, DC: Gales and Seaton, 1856), 18th Cong., 1st sess., 2:3217.

16. Robert G. Angevine, "Individuals, Organizations, and Engineering: U.S. Army Officers and the American Railroads, 1827–1838," *Technology and Culture* 42.2 (April 2001): 298. Philip E. Thomas was the first president of the B&O Railroad.

17. Annual Report of the Secretary of War, 24 November 1828, *American State Papers, Military Affairs*, 4:2.

18. Thayer to Macomb, 14 December 1825, quoted in Crackel, 97.

19. David B. Douglass to Sylvanus Thayer, 12 January 1826, in *The West Point Thayer Papers, 1802–1872*, section 4.

20. Ticknor to his wife, 17 June 1826, in George Ticknor, *Life, Letters, and Journals of George Ticknor* (Boston: James R. Osgood, 1877), 1:375.

21. For examples, consider the careers of Civil War veterans from West Point. George B. McClellan, Robert E. Lee, Philip Sheridan, William T. Sherman, and Ulysses S. Grant are all representative of "military professionalism" as first exemplified by Thayer.

22. Leeman, 78.

23. USMA Association of Graduates, *Register of Graduates and Former Cadets of the United States Military Academy, Class of 1900 Centennial Edition* (West Point, NY, 2000), 4–25.

24. Board of Visitors, *Report of the Board of Visitors* (24 June 1826), 16.

25. Thayer to Swift, 29 February 1832, in *The West Point Thayer Papers, 1802–1872*, section 5. Thayer wrote: "The only source of hope is in the President (Andrew Jackson) who is in the habit of dispensing with the most important regulations of the Academy in favor of his friends in spite of the Academic Authorities & the Secretary of War himself. I do not see why he will not be as likely to yield to any solicitations from the friends of Cadet Adams as to those of others in behalf of their relations. The chances of success would, in my opinion, be as 3 to 1."

26. Crackel, 101–102.

27. Ticknor to George W. Cullum, 29 May 1864, in *The West Point Thayer Papers, 1802–1872*, section 10.

28. Crackel, 81, 101.

29. James William Kershner, *Sylvanus Thayer: A Biography* (New York: Arno, 1982), 314–15, 323.

30. Ambrose, *Duty, Honor, Country,* 189; Horace Webster to Sylvanus Thayer, 19 May 1869, in *The West Point Thayer Papers, 1802–1872,* section 10.

31. Thomas E. Griess, "Dennis Hart Mahan: West Point Professor and Advocate of Military Professionalism, 1830–1871" (PhD diss., Duke University, 1968), 178, 347.

32. Alfred Thayer Mahan, *From Sail to Steam: Recollections of Naval Life* (New York: Harper & Brothers, 1907), ix.

33. Ibid.; Henry L. Abbott, "Memoir of Dennis Hart Mahan: Read before the National Academy, November 7, 1878," 31.

34. Suzanne Geissler, "Professor Dennis Mahan Speaks Out on West Point Chapel Issues, 1850," *Journal of Military History* 69.2 (April 2005): 505–19, 508.

35. Dennis Hart Mahan to Esther M. Mahan, 29 June 1824, Alfred Thayer Mahan Papers, Library of Congress (microfiche).

36. Geissler, 514–15.

37. John F. Marszalek, *Commander of All Lincoln's Armies: A Life of General Henry W. Halleck* (Cambridge, MA: Belknap Press of Harvard University Press, 2004), 22.

38. Dennis Hart Mahan, *An Elementary Treatise on Advanced Guard, Out Post, and Detachment Service of Troops* (New York: John Wiley, 1863), 266.

39. Brian M. Linn, *The Echo of Battle: The Army's Way of War* (Cambridge, MA: Harvard University Press, 2007), 23. Among the most famous nineteenth-century military theorists, Carl von Clausewitz and Antoine-Henri Jomini both base their theories on the timeless aspects of war.

40. William B. Skelton, *An American Profession of Arms: The Army Officer Corps, 1784–1861* (Lawrence: University Press of Kansas, 1992), 168. As a biographical note, when Mahan became professor of engineering in 1832, he "formally vacated his commission in the corps of engineers to accept the professorship of civil and military engineering." See Abbott, 32.

41. Marszalek, 27; Ambrose *Duty, Honor, Country,* 136–37.

42. Dennis Hart Mahan, *A Treatise on Field Fortification, Containing Instructions on the Methods of Laying Out, Constructing, Defending, and Attacking Intrenchments, with the General Outlines Also of The Arrangement, The Attack and Defence of Permanent Fortifications* (New York: John Wiley, 1836, 1862), 46.

43. Griess, 173.

44. Dennis Hart Mahan to Frederick Harris, 5 April 1834, in *The Education of Col. David Bullock Harris, C.S.A.: West Point Letters (1829–'35),* ed. Charles W. Turner (Verona, VA: McClure, 1984), 98. David Bullock Harris resigned from his commission in 1835 and returned to his native Virginia to be a planter and engineer. During the Civil War, he served as an engineer in the Confederate States Army, achieving brigadier general before dying in Charleston, South Carolina, 10 October 1864.

45. Griess, 178, 181, 196, 197, 224.

46. David M. Reel, "The Drawing Curriculum at the U.S. Military Academy during the 19th Century," in *West Point Points West* (Denver: Institute of Western American Art, 2002), 52.

47. William Cullen Bryant to Lewis Cass, 19 March 1834, in *The Letters of William Cullen*

Bryant, 1809–1836, ed. William Cullen Bryant II and Thomas G. Voss (New York: Fordham University Press, 1975), 1:395–96.

48. George W. Cullum, *Biographical Register of the Officers and Graduates of the U.S. Military Academy at West Point* (Cambridge, MA: Riverside Press, 1891), 1:39.

49. Reel, 55–56.

50. See Seth Eastman, *Treatise on Topographical Drawing* (New York: Wiley and Putnam, 1837). Eastman's textbook of over one hundred pages includes descriptions and examples of terrain and topographical symbols that are seen in many of the topographical maps and drawings between 1840 and 1900.

51. Morrison, *"The Best School,"* 97.

52. Reel, 56.

53. Marvin J. Anderson, "The Architectural Education of Nineteenth-Century American Engineers: Dennis Hart Mahan at West Point," *Journal of Society of Architectural Historians* 67.2 (2008): 222, 227, 229–31.

54. Griess, 185.

55. Marszalek, 27; United States Military Academy, *Regulations of the U.S. Military Academy at West Point* (New York: J&J Harper, 1832), 10–14.

56. Griess, 190; Samuel B. McIntire, "Echoes of the Past," *Army and Navy Journal* 39 (14 June 1902): 1026.

57. Griess, 173, 174, 181.

58. Calhoun, 82–83.

59. USMA Staff Records, 3 (1842–45), 195–96.

60. Griess, 243.

61. Russell F. Weigley, "American Strategy from Its Beginnings through the First World War," in *Makers of Modern Strategy: From Machiavelli to the Nuclear Age* (Princeton, NJ: Princeton University Press, 1986), 414–15.

62. Abbott, 33.

63. USMA Association of Graduates, *Annual Reunion of the Association of Graduates of the United States Military Academy,* 14 June 1872, 24.

64. Abbott, 29–37; *Annual Reunion* (1872), 18–25.

65. Griess, 206–207, 346.

66. Molloy, 459–60.

67. Griess, 202–203, 204; *Annual Reunion* (1872), 20; Abbott, 34.

68. Calhoun, 208.

79. Molloy, 459–60.

70. Manning, 246–47.

71. William H. Wisely, *The American Civil Engineer, 1852–1974: The History, Traditions, and Development of the American Society of Civil Engineers* (New York: American Society of Civil Engineers, 1974), 4, 77.

72. R. K. McMaster, *West Point's Contribution to Education, 1802–1952* (New York: McMath, 1952) 9.

73. Abbott, 31.

74. Thayer to President Asa Smith (Dartmouth College President), 26 July 1867, in Dennis Hart Mahan Papers, MSS Collections, Dartmouth College Library, Hanover, NH, in Griess, 346.

75. Crackel, 134–35.

76. Board of Visitors, *Report of the Board of Visitors* (June 1867), 195.

4. ENGINEERING PROFESSIONALS IN NEW YORK: A NEW AMERICAN IDENTITY

1. Edwin G. Burrows and Mike Wallace, *Gotham: A History of New York City to 1898* (New York: Oxford University Press, 1999), 430.

2. Charles Sellers, *The Market Revolution: Jacksonian America, 1815–1846* (New York: Oxford University Press, 1991), 42–43.

3. Edward K. Spann, *The New Metropolis: New York City, 1840–1857* (New York: Columbia University Press, 1981), 5.

4. Robert G. Albion, *The Rise of New York Port, 1815–1860* (New York: Charles Scribner's Sons, 1939), 12–14, 60–61, 90–94, 119, 276–79.

5. Ibid., 55–60; George J. Lankevich, *New York City: A Short History* (New York: New York University Press, 2002), 60–68.

6. Albion, 378–85.

7. William Cronon, *Nature's Metropolis: Chicago and the Great West* (New York: Norton, 1991), 60.

8. Peter L. Bernstein, *Wedding of the Waters: The Erie Canal and the Making of a Great Nation* (New York: Norton, 2005), 27.

9. Spann, *The New Metropolis*, 46.

10. Sellers, 40–41.

11. Gerard T. Koeppel, *Water for Gotham: A History* (Princeton, NJ: Princeton University Press, 2000), 158; Larry D. Lankton, *The "Practicable" Engineer: John B. Jervis and the Old Croton Aqueduct* (Chicago: Public Works Historical Society, 1977), 4.

12. Koeppel, 172; Lankton, 4.

13. George S. Greene, Handwritten Record of G. S. Greene, 1801–1897, George Sears Greene Collection, Rhode Island Historical Society MSS 460, Box 1, Folder 6.

14. Jon C. Teaford, *The Unheralded Triumph: City Government in America, 1870–1900* (Baltimore: Johns Hopkins University Press, 1984), 136–37.

15. Robert V. Bruce, *The Launching of Modern American Science, 1846–1876* (Ithaca, NY: Cornell University Press, 1987), 341–42.

16. Burton J. Bledstein, *The Culture of Professionalism: The Middle Class and the Development of Higher Education in America* (New York: Norton, 1978), x.

17. Ibid., 8–24, 31.

18. Jennifer R. Green, *Military Education and the Emerging Middle Class in the Old South* (New York: Cambridge University Press, 2008), 15–16. Thompson writes: "If we stop history at a given point, then there are no classes but simply a multitude of experiences. But if we watch these men

over an adequate period of social change, we observe patterns in their relationships, their ideas, and their institutions. Class is defined by men as they live their own history, and, in the end, this is its only definition." See E. P. Thompson, *The Making of the English Working Class* (New York: Vintage, 1966), 11.

19. Sellers, 237–38.

20. Stuart M. Blumin, *The Emergence of the Middle Class: Social Experience in the American City, 1760–1900* (New York: Cambridge University Press, 1989), 1–2.

21. Sven Beckert, *The Monied Metropolis: New York and the Consolidation of the American Bourgeoisie, 1850–1896* (New York: Cambridge University Press, 2001), 6–8.

22. Robert H. Wiebe, *The Search for Order, 1877–1920* (New York: Hill and Wang, 1967), 111–32.

23. James L. Morrison Jr., *"The Best School": West Point, 1833–1866* (Kent, OH: Kent State University Press, 1998), 62.

24. Bledstein, 5.

25. Morrison, *"The Best School,"* 61, 155.

26. William B. Skelton, *An American Profession of Arms: The Army Officer Corps, 1784–1861* (Lawrence: University Press of Kansas, 1992), 165–67.

27. Bledstein, 195.

28. USMA Staff Records 3 (1842–45), 195–96.

29. Bledstein, 89. For Williams, see Peter Michael Molloy, "Technical Education and the Young Republic: West Point as America's Ecole Polytechnique, 1802–1833" (PhD diss., Brown University, 1975), 241–42.

30. Beckert, 8.

31. William H. Wisely, *The American Civil Engineer, 1852–1874: The History, Traditions, and Development of the American Society of Civil Engineers* (New York: American Society of Civil Engineers, 1974), 13. The Rotunda Park was at Chambers Street on the Lower East Side, in what today is known as City Hall Park. See New York City Department of Parks & Recreation website, accessed 13 October 2010 at http://www.nycgovparks.org/parks/cityhallpark/.

32. Charles Warren Hunt, *Historical Sketch of the American Society of Civil Engineers* (New York, 1897), 9–12. In his 1974 history of the ASCE, William H. Wisely estimates that these forty men were about 10 percent of the known civil engineers in the United States. See Wisely, 7.

33. James D. Dilts, *The Great Road: The Building of the Baltimore and Ohio, The Nation's First Railroad, 1828–1853* (Stanford, CA: Stanford University Press, 1993), 3.

34. Ibid., 63; USMA Association of Graduates, *Register of Graduates and Former Cadets of the United States Military Academy, Class of 1900 Centennial Edition* (West Point, NY, 2000). The USMA graduates were Joshua Barney (1820), Isaac Ridgeway Trimble (1822), Walter Gwynn (1822), William Cook (1822), John N. Dillahunty (1824), R. Edward Hazzard (1824), John M. Fessenden (1824), and William B. Thompson (1824). Walter B. Guion spent three years as a cadet before resigning in 1827. Frederick Harrison Jr. had been admitted in 1823 with the class of 1827. Knight was a "civil, nonmilitary, engineer, although often employed by the federal government." Wever could be classified as a "practical engineer and builder." See also Robert G. Angevine, "Individuals, Organizations, and Engineering: U.S. Army Officers and the American Railroads, 1827–1838," *Technology and Culture* 42.2 (April 2001): 298.

35. *Debates and Proceedings in the Congress of the United States, 1834–1856* (Washington, DC: Gales & Seaton, 1856), 18th Cong., 1st sess., 2:3217.

36. Forest G. Hill, *Roads, Rails, & Waterways: The Army Engineers and Early Transportation* (Norman: University of Oklahoma Press, 1957), 100–101.

37. Hunt, *Historical Sketch*, 10–11; USMA Association of Graduates, *Register of Graduates and Former Cadets of the United States Military Academy, Class of 1900 Centennial Edition* (West Point, NY, 2000).

38. Hunt, *Historical Sketch*, 12; Samuel Smiles, *The Life of Thomas Telford, Civil Engineer: With an Introductory History of Roads and Travelling in Great Britain* (London: John Murray, Albemarle Street, 1867), 304–305.

39. Bledstein, 193; Raymond H. Merritt, *Engineering in American Society, 1850–1875* (Lexington: University Press of Kentucky, 1969), 1–26.

40. Hunt, *Historical Sketch*, 10–13. The Committee of Seventeen were Benjamin Wright of New York, William S. Campbell of Florida, Claude Crozet of Virginia, William C. Fairfax of Virginia, C. B. Fisk of Maryland, Edward F. Gay of Pennsylvania, Walter Gwynn of North Carolina, J. B. Jervis of New York, Jonathan Knight of Maryland, Benjamin Latrobe of Maryland, W. G. McNeill of South Carolina, Edward Miller of Pennsylvania, Moncure Robinson of Virginia, J. Edgar Thomson of Georgia, Isaac Trimble of Maryland, Sylvester Welch of Kentucky, and G. W. Whistler of Connecticut.

41. Wisely, 12–13.

42. Koeppel, 140–43.

43. Charles King, *A Memoir of the Construction, Cost, and Capacity of the Croton Aqueduct, Compiled from Official Documents* (New York: Charles King, 1843), 113.

44. Koeppel, 150–53.

45. "Report of Mr. D. B. Douglass to the Commissioners for Supplying the City of New York with Pure and Wholesome Water," dated 1 February 1835, in Document No. 44, *Documents, Board of Aldermen of the City of New York*, 1: 403–33 (1835).

46. Koeppel, 153, 156–59. In the nineteenth century, death and disease often played a pivotal role in who became successful and who did not. The sickness and death of Clinton in 1833–1834 and the death of Canvass White in 1833 made Douglass the best-known available, living engineer to lead the project.

47. Edward Wegmann, *The Water-Supply of the City of New York, 1658–1895* (New York: John Wiley & Sons, 1896), 37; Lankton, 8.

48. Koeppel, 160, 172.

49. Wegmann, 37.

50. Koeppel, 186–89.

51. Lankton, 3.

52. Wegmann, 45.

53. Koeppel, 280–81, 287; Wegmann, 48. In his report of 27 July 1842, Jervis estimated the cost to be "nearly" $9 million. See John B. Jervis, *Description of the Croton Aqueduct* (New York: Slamm and Guion, 1842), 31.

54. Koeppel, 283.

55. Wegmann, 61.

56. *The National Cyclopedia of American Biography* 9:37 (1907).

57. Wegmann, 239.

58. "Alfred W. Craven," *New York Times,* 29 March 1879, 5.

59. George S. Greene to Alfred W. Craven, 6 May 1841, in Craven Family Papers, New-York Historical Society; David W. Palmer, *The Forgotten Hero of Gettysburg: A Biography of General George Sears Greene* (Philadelphia: Xlibris, 2005), 37–38.

60. Francis Vinton Greene, "Memoir of the Life and Services of George Sears Greene" (New York, 1902), 5.

61. Ibid., 2–3; Palmer, 22–23.

62. Greene, "Memoir of the Life and Services," 4. Greene was commissioned as an officer in the artillery branch. In most classes, only the top one or two cadets were commissioned into the Corps of Engineers.

63. Ibid., 5. There are several biographical sources for G. S. Greene, but the memoirs by Greene's son, Francis, are the best summary of the existing records.

64. Palmer, 47.

65. American Society of Civil Engineers, *Proceedings of the American Society of Civil Engineers,* Nov. 1873–Dec. 1875 (New York: ASCE, 1876), 1:41; USMA, *Official Register of the Officers and Cadets of the United States Military Academy* (1833), 7.

66. "Report of Mr. D. B. Douglass to the Commissioners," 418.

67. George Cullum, *Biographical Register of the Officers and Graduates of the Military Academy at West Point, N. Y.,* 3rd ed. (New York: Houghton, Mifflin, 1891), 1:541; USMA Association of Graduates, *Annual Reunion of the Association of Graduates of the United States Military Academy,* 1874, 8.

68. Cullum, 541.

69. American Society of Civil Engineers, *Proceedings,* 1:42; Cullum, 541.

70. Alexander Hamilton Jr. to his Grandfather Robert Morris, 27 April 1843, in Misc. MSS, Alexander Hamilton Jr., Letters 1823–1882, New-York Historical Society, Folder 1.

71. "Account of the Croton Aqueduct for Supplying New York with Water," *Mechanic's Magazine, Museum, Register, Journal, and Gazette* 32 (12 October 1839): 24.

72. Burrows and Wallace, xxi, 655; Wisely, 19.

73. Burrows and Wallace, 908.

74. Spann, *The New Metropolis,* 7, 10.

75. Rep. Willis Gorman, 4 February 1851, *Congressional Globe,* 31st Congress, 2nd Sess., p. 417.

76. Charles King, 116, 143.

77. Mark Aldrich, "Earnings of American Civil Engineers, 1820–1859," *Journal of Economic History* 31.2 (June 1971): 412.

78. Board of Aldermen, *Annual Report of the Croton Aqueduct Department Made to the Common Council of the City of New York,* 3 January 1855 (New York, 1855), 31.

79. Daniel H. Calhoun, *The American Civil Engineer: Origins and Conflict* (Cambridge, MA: Harvard University Press, 1960), 167–69, 207.

80. Wisely, 5.

81. Hunt, *Historical Sketch*, 16. The invitation read:

Dear Sir:

A meeting will be held at the office of the Croton Aqueduct Department, Rotunda Park on Friday, November 5th, at 7 o'clock, P.M., for the purpose of making arrangements for the organization, in the city of New York, of a Society of Civil Engineers and Architects.

Should the object of the meeting obtain your approval, you are respectfully invited to attend.

[signed] Wm. H. Morell, Wm. H. Sidell, J. W. Adams, A. W. Craven, James Laurie, James P. Kirkwood, and others

82. "Col. Julius W. Adams Dead," *New York Times,* 14 December 1899, 14; American Society of Civil Engineers biography, accessed 6 November 2015 at http://www.asce.org/templates/person-bio-detail.aspx?id=9819 . . ./; Francis Gbondo, "Engineering History: Julius Walker Adams (1812–1899)," *American Society of Civil Engineers San Bernardino & Riverside Counties Branch Newsletter,* November 2009, 2. Adams apparently left West Point after one year to work as an assistant engineer with his uncle, Maj. George W. Whistler, on the Paterson and Hudson Railroad. After the Civil War, Adams worked under Washington Roebling on the Brooklyn Bridge (see chapter 6).

83. Hunt, *Historical Sketch,* 20.

84. Ibid., 43. The name American Society of Civil Engineers was officially registered in 1877.

85. Charles Warren Hunt, "Address at the Annual Convention at Washington, D.C., May 20th, 1902: The First Fifty Years of the American Society of Civil Engineers: 1852–1902," *Transactions of the American Society of Civil Engineers* 48 (New York, 1902).

86. Ibid., 221–22.

87. Wisely, 20–21, 77.

88. Hunt, *Historical Sketch,* 24–25.

89. Wisely, 106.

90. Alfred W. Craven, "President's Address to the Annual Convention of the American Society of Civil Engineers, 1870," in Wisely, 28.

91. Wisely, 21–23.

92. Dell Upton, *Another City: Urban Life and Urban Spaces in the New American Republic* (New Haven, CT: Yale University Press, 2008), 1.

93. Palmer, 44–45.

94. American Society of Civil Engineers, *Transactions,* 49:335–36.

95. Palmer, 48; American Society of Civil Engineers, *Transactions,* 49:336.

96. Wegmann, 66; Frederick Law Olmsted, "Description of the Central Park, January, 1859," in *The Papers of Frederick Law Olmsted,* ed. Charles E. Beveridge and David Schuyler, vol. 3, *Creating Central Park, 1857–1861* (Baltimore: Johns Hopkins University Press, 1983), 205.

97. Joanne Abel Goldman, *Building New York's Sewers: Developing Mechanisms of Urban Management* (West Lafayette, IN: Purdue University Press, 1997), 122–23.

98. Wegmann, 66.

99. Ibid.; "The New Reservoir," *New York Times,* 19 April 1858, 5.

100. Board of Commissioners of the Central Park, *Fourth Annual Report* (New York: Wm. C. Bryant, 1861), 110.

101. Evan Cornog, "Whig Party," in *The Encyclopedia of New York City*, ed. Kenneth T. Jackson (New Haven, CT: Yale University Press, 1995), 1257.

102. Greene, "Memoir of the Life and Services," 5–6.

103. See James M. McPherson, *Battle Cry of Freedom: The Civil War Era* (New York: Oxford University Press, 1988), 660–61; John Cox, *Culp's Hill: The Attack and Defense of The Union Flank, July 2, 1863 (Battleground America)* (Cambridge, MA: Da Capo, 2003), 50–173; and David W. Palmer, *The Forgotten Hero of Gettysburg: A Biography of General George Sears Greene*, 145–208.

104. Major General H. W. Slocum to Major General George G. Meade, 30 December 1863, in George Sears Greene Collection, Rhode Island Historical Society MSS, Box 1, Folder 2.

105. Greene, "Memoir of the Life and Services," 9–12.

106. Martin V. Melosi, *The Sanitary City: Environmental Services in Urban America from Colonial Times to the Present* (Pittsburgh: University of Pittsburgh Press, 2008), 45.

107. Citizens' Association of New York, *Report of the Council of Hygiene and Public Health of the Citizens' Association of New York upon the Sanitary Condition of the City*, 2nd ed. (New York: D. Appleton, 1866), cxlii–cxlii.

108. American Society of Civil Engineers, *Transactions*, 49:337.

109. Wegmann, 79; American Society of Civil Engineers, *Transactions*, 49:337.

110. Ibid.

111. George S. Greene, Handwritten Record of G. S. Greene, 1801–1897, George Sears Greene Collection, Rhode Island Historical Society MSS 460, Box 1, Folder 6.

112. Goldman, 146–47.

113. Goldman, 149–55; Burrows and Wallace, 928.

114. Burrows and Wallace, 1010.

115. American Society of Civil Engineers, *Transactions*, 49:338.

116. George Sears Greene Collection, Rhode Island Historical Society MSS (Memorandum Book of Engineering Calculations, June 1874), Box 1, Folder 9.

117. Skelton, 220.

118. "Gen. George S. Greene Dead," *New York Times*, 28 January 1899.

119. Hunt, *Historical Sketch*, 33.

120. Wisely, 25.

121. Hunt, *Historical Sketch* 34.

122. Wisely, 29.

123. Burrows and Wallace, 969; Wisely, 100.

124. Beckert, 115. Also see Stanley K. Schultz and Clay McShane, "To Engineer the Metropolis: Sewers, Sanitation, and City Planning in Late-Nineteenth-Century America," *Journal of American History* 65.2 (September 1978): 401.

125. "The American Society of Civil Engineers," *Engineering News and American Contract Journal*, vol. 17, January–June 1887 (New York: Engineering News, 1887), 79.

126. "The Croton Aqueduct Investigation Second Meeting of the Aldermanic Committee," *New York Herald*, 12 September 1860, 4.

127. Board of Aldermen, "Document 18: Report of the Special Committee of the Board of

Aldermen, Appointed to Investigate the Sufficiency of the Charges Alleged by His Honor the Mayor for the Removal of Messrs. Craven and Tappen," in *Documents of the Board of Aldermen of the City of New York,* vol. 27, part 2 (New York: Edmund Jones, 1860), 57.

128. Ibid., 36.

129. Ibid., 40; Alfred W. Craven, "President's Address to the Annual Convention of the American Society of Civil Engineers, 1870," in Wisely, 28. In his address, Craven praised George S. Greene by writing, "A more thorough investigator or more upright man can nowhere be found."

130. Teaford, 137.

131. American Society of Civil Engineers, *Proceedings of the American Society of Civil Engineers,* Jan.–Dec. 1880 (New York: ASCE, 1880), 6:25–26.

132. American Society of Civil Engineers, *Transactions,* 49:340.

133. Greene, "Memoir of the Life and Services," 13.

134. George S. Greene, Handwritten Record of G. S. Greene, 1801–1897, George Sears Greene Collection, Rhode Island Historical Society MSS 460, Box 1, Folder 6; Francis Vinton Greene, "Memoir of the Life and Services," 13.

135. Wisely, 30. His son, George S. Greene Jr., as chief engineer of the docks, would be implicated in an investigation of Tammany's Richard Croker in 1895, when the younger Greene admitted to paying off the Tammany boss $125 in 1892 and 1893. See Teaford, 137.

136. American Society of Civil Engineers, *Transactions,* 49:339–40.

137. Hunt, *Historical Sketch,* 79.

138. American Society of Civil Engineers, *Transactions,* 49:339.

139. *Annual Reunion* (11 June 1888), 11.

140. Beckert, 253–54.

5. EGBERT L. VIELE'S NEW YORK

1. Charles E. Beveridge and David Schuyler, eds., *The Papers of Frederick Law Olmsted,* vol. 3, *Creating Central Park, 1857–1861* (Baltimore: Johns Hopkins University Press, 1983), 87 (hereafter abbreviated to *PFLO* 3); Jerome Mushkat, *Fernando Wood: A Political Biography* (Kent, OH: Kent State University Press, 1990), 43, 47.

2. *PFLO* 3:87, 93n16.

3. Frederick Law Olmsted, "Passages in the Life of an Unpractical Man," in *PFLO* 3:89; Justin Martin, *Genius of Place: The Life of Frederick Law Olmsted* (Cambridge, MA: Da Capo, 2011), 130–31.

4. *PFLO* 3:88, 98n17.

5. Iver Bernstein, *The New York City Draft Riots: Their Significance for American Society and Politics in the Age of the Civil War* (New York: Oxford University Press, 1990), 43–71. Bernstein's chapter 2, "The Two Tempers of Draco," is one of the best descriptions of the two separate (radical vs. conservative) visions for reform in New York City.

6. Anthony Gronowicz, *Race and Class Politics in New York City before the Civil War* (Boston: Northeastern University Press, 1998), 129–51; George J. Lankevich, *New York City: A Short History* (New York: New York University Press, 2002), 169.

7. Edwin G. Burrows and Mike Wallace, *Gotham: A History of New York City to 1898* (New York: Oxford University Press, 1999), 930–31. For a depiction of the difference between East Side and West Side water and sewer infrastructure after the Civil War, see the maps in Eugene P. Moehring, *Public Works and the Patterns of Urban Real Estate Growth in Manhattan, 1835–1894* (New York: Arno, 1981), 320–21.

8. See Egbert L. Viele, "Report on the Civic Cleanliness, and the Economical Disposition of the Refuse of Cities" (New York: Edmund Jones, 1860), 34–36; *Real Estate Record and Builders' Guide* 28 (26 November 1881); and Roy Rosenzweig and Elizabeth Blackmar, *The Park and the People: A History of Central Park* (Ithaca, NY: Cornell University Press, 1992), 100.

9. "Egbert L. Viele," in *Representative Men of New York: A Record of Their Achievements,* ed. Jay Henry Mowbray (New York: New York Press, 1898), 2:163–65. Viele's memberships included the London Club, the Association of Graduates of West Point, the Grand Army of the Republic, the Loyal Legion, the Century Club, the New York Club, the Union League, the Saint Nicholas Society, the Aztec Club, the Holland Society, the National Academy, the Geographical Society, and the Genealogical Society.

10. USMA Association of Graduates, *Annual Reunion of the Association of Graduates of the United States Military Academy,* 1902, 142–44; Chase Viele, "America's Pyramid on-the-Hudson," *Assembly* (December 1973): 20; Jon Scott Logel, "Party, Park, and a Pyramid: Egbert L. Viele and the Creation of Central Park," *De Halve Maen* 75.4 (Winter 2002): 63–68.

11. James R. Endler, *Other Leaders, Other Heroes: West Point's Legacy to America beyond the Field of Battle* (Westport, CT: Praeger, 1998), 68.

12. Daniel M. Bluestone, "From Promenade to Park: The Gregarious Origins of Brooklyn's Park Movement," *American Quarterly* 39.4 (Winter 1987): 542.

13. Rosenzweig and Blackmar, 6–7.

14. William Cullen Bryant, "A New Public Park," *Evening Post,* 42, 3 July 1844, 2.

15. Rosenzweig and Blackmar, 16–20, 44–46, 59, 97. Jones Wood was to run between 66th and 75th Streets. As noted in chapter 4, the Croton Aqueduct and Water System cost well over $12 million by 1842.

16. "New-York City: Commissioners of Central Park," *New York Times,* 6 June 1856, 3; "A Good Beginning," *New York Times,* 10 June 1856, 4; "New-York City: The Present Look of Our Great Central Park," *New York Times,* 9 July 1856, 3. Between 10 June and 9 July 1856, the commissioners named Viele engineer-in-chief.

17. Chase Viele, "The Knickerbockers of Upstate New York," *De Halve Maen* 47 (October 1972): 1–2.

18. George W. Cullum, *Biographical Register of the Officers and Graduates of the United States Military Academy,* vol. 2, 3rd ed. (New York: Houghton Mifflin, 1891), no. 1360, 338.

19. USMA, *Official Register of the Officers and Cadets of the United States Military Academy* (1838–1847), 11, 21.

20. George S. Pappas, *To the Point: The United States Military Academy, 1802–1902* (Westport, CT: Praeger, 1993), 248.

21. *The National Cyclopaedia of American Biography* (New York: James T. White, 1921), 2:195.

22. Ibid.; Cullum, *Biographical Register of the Officers and Graduates of the United States Military Academy,* no. 1360.

23. George Templeton Strong, *The Diary of George Templeton Strong*, vol. 2, *The Turbulent Fifties, 1850–1859*, ed. Allan Nevins and Milton Halsey Thomas (New York: Macmillan, 1952), xi, 83.

24. *Dictionary of American Biography*, 10:267; Strong, *The Diary of George Templeton Strong*, vol. 4, *Post-War Years, 1865–1875*, ed. Nevins and Thomas, 314. The New York press followed the divorce closely. Highlights of this public case include allegations of adultery by both Egbert and Teresa. According to George Templeton Strong, General Viele accused Teresa of having an affair with Gen. William Averell, a West Point classmate of Viele's. Teresa accused Egbert of carrying on with Julia Dana, who did become Viele's second wife. For press reports, see "The Viele Scandal," *New York Times*, 18 June 1871, 8; "The Viele Scandal," *New York Times*, 21 June 1871, 5; "The Viele Scandal Custody of the Children," *New York Times*, 22 June 1871, 3; "The Viele Divorce," *New York Daily News*, 30 August 1871, 1; "Viele Divorce," *New York Sun*, 2 September 1871, 1; "The Viele Kidnapping Case," *New York Times*, 19 September 1872, 1; "The Custody of the Viele Children More Complex Developments in the Case," *New York Times*, 20 September 1872, 2.

25. *The National Cyclopaedia of American Biography*, 2:195.

26. Egbert L. Viele, "Report," 2.

27. Oliver E. Allen, *The Tiger: The Rise and Fall of Tammany Hall* (New York: Addison-Wesley, 1993), 73; Mushkat, 39; James F. Richardson, *The New York Police: Colonial Times to 1901* (New York: Oxford University Press, 1970), 98–103.

28. Rosenzweig and Blackmar, 97. The eleven commissioners were Robert J. Dillon, James E. Cooley, Charles H. Russell, John F. Butterworth, John A. C. Gray, Waldo Hutchins, Thomas E. Field, Andrew H. Greene, Charles W. Elliott, William K. Strong, and James Hogg. Also see *Documents of the Board of Commissioners of the Central Park for the Year Ending April 30, 1858* (1858), 10; and Mushkat, 69.

29. Allen, 73.

30. Rosenzweig and Blackmar, 100–101.

31. Paul E. Cohen and Robert T. Augustyn, *Manhattan in Maps, 1527–1995* (New York: Rizzoli International, 1997), 130.

32. Egbert L. Viele, "Report," 12. Working under Viele in 1856–1857 was George Waring, the great sanitation advocate and future New York City street commissioner. Also see Rosenzweig and Blackmar, 162–63.

33. Egbert L. Viele, "Report," 13.

34. Cohen and Augustyn, 130.

35. Rosenzweig and Blackmar, 65–73. In 2011, a team of anthropologists, archaeologists, and student interns with the Institute for the Exploration of Seneca Village History conducted an archaeological dig on the site, further expanding the research first highlighted by Rosenzweig and Blackmar. See Lisa W. Foderaro, "Unearthing Traces of African-American Village Displaced by Central Park," *New York Times*, 27 July 2011, accessed 26 August 2011 at http://www.learn.columbia.edu/seneca_village/index.html.

36. Egbert L. Viele, *Map of the Lands Included in the Central Park from a Topographical Survey, June 17th 1856*, New-York Historical Society; Rosenzweig and Blackmar, 64–65.

37. Catherine McNeur, *Taming Manhattan: Environmental Battles in the Antebellum City* (Cambridge, MA: Havard University Press, 2014), 203–205; Egbert L. Viele, "Topography of New-

York and Its Park System," in *The Memorial History of the City of New-York and the Hudson River Valley*, ed. James Grant Wilson (New York: New-York History, 1892), 4:556–57.

38. Egbert L. Viele, "Report," 11, 36–37, 39–40, 42. As described in chapter 4, George S. Greene designed and built the supporting pump structures for the "Croton Lake." See Edward Wegmann, *The Water-Supply of the City of New York, 1658–1895* (New York: John Wiley & Sons, 1896), 66.

39. Rosenzweig and Blackmar, 101–102. Landscape architects were one of the later professions to organize in the nineteenth century, forming the American Society of Landscape Architects in 1899. Frederick Law Olmsted led the profession in defining its expertise and authority. Unlike civil engineering, landscape architecture in the nineteenth century remained largely a body of expertise gained through experience and not formal education. See Albert Fein, *Landscape into Cityscape: Frederick Law Olmsted's Plans for a Greater New York City* (Ithaca, NY: Cornell University Press, 1968), 1–42, 385; Frederick Law Olmsted Sr., *Forty Years of Landscape Architecture: Central Park*, ed. Frederick Law Olmsted Jr. and Theodora Kimball (Cambridge, MA: MIT Press, 1973), viii–ix.

40. Witold Rybczynski, *A Clearing in the Distance: Frederick Law Olmsted and America in the Nineteenth Century* (New York: Scribner, 1999), 161–62.

41. Rosenzweig and Blackmar, 124.

42. Francis R. Kowsky, *Country, Park, and City: The Architecture and Life of Calvert Vaux* (New York: Oxford University Press, 1998), 96–97; Martin, 139–42.

43. *PFLO* 3:117–77.

44. Kowsky, 97–98.

45. Cohen and Augustyn, 134.

46. *PFLO* 3:90.

47. Olmsted to Dr. Asa Gray, 8 October 1857, in *PFLO* 3:102.

48. Olmsted to John Olmsted (his father), 9 October 1857, in *PFLO* 3:104 and 105n1. In that same letter, Olmsted also asked his father in London to get a "nice, thin, light, silk faced, English India rubber [over-coat]" for "Colonel Viele."

49. Rosenzweig and Blackmar, 119, 553–54n38. The authors' table of plans submitted show at least twelve of the thirty-three entries as coming from park employees. No. 28 is credited to Viele, but interestingly enough, submission no. 2 is an anonymously signed pyramid.

50. Board of Commissioners of the Central Park, *Minutes*, 17 May 1858, 30–31, cited in *PFLO* 3:192.

51. *PFLO* 3:198–99n2; *Morning Courier & New-York Enquirer*, 31 May 1858, 2; "The Central Park Job," *New York Herald*, 31 May 1858, 4; *New York Daily Tribune*, 31 May 1858, 4.

52. *Viele v. Gray*, Supreme Court, New York County, New York, December term 1859, in West Headnotes: 1859 WL 7727 (N.Y. Com.Pl.).

53. Rosenzweig and Blackmar, 374.

54. Olmsted, *Forty Years*, 558, 560–61.

55. Olmsted to Vaux, 25 March 1865, in Victoria Post Ranney, ed., *The Papers of Frederick Law Olmsted*, vol. 5, *The California Frontier, 1863–1865* (Baltimore: Johns Hopkins University Press, 1990), 209–11.

56. Olmsted, *Forty Years*, 557, 560–61.

57. Kowsky, 165–66. Also see Suellen Hoy, *Chasing Dirt: The American Pursuit of Cleanliness*

(New York: Oxford University Press, 1995), 34–38. Bellows was a Unitarian minister who collaborated with Elizabeth Blackwell and prominent New York physicians to create the Women's Central Association of Relief, the forerunner to the Sanitary Commission, in 1861. By 1864, the time of the trial, Bellows and Olmsted had formed a tighter relationship through their Sanitary Commission efforts and shared Republican beliefs.

58. Egbert L. Viele, "Topography of New-York and Its Park System," 4:557.

59. Fremont Rider, *Rider's New York City* (New York, 1916), 301, as quoted in Chase Viele, "A Short Biography of Egbert L. Viele" (unpublished manuscript, USMA Special Collections, 1973), 34. Rider's use of the term *landscape gardeners* in a 1916 publication highlights the early challenges of landscape architects at the turn of the twentieth century.

60. Henry Hope Reed and Sophia Duckworth, *Central Park: A History and a Guide* (New York: Clarkson N. Potter, 1967), 17–19; Rosenzweig and Blackmar, 199–205.

61. "Hands off Fifty-Ninth Street," *New York Times,* 24 June 1879, 8; "Park Reforms Suggested," *New York Times,* 8 February 1883, 3; "Asking for Improvement," *New York Times,* 14 June 1885, 2; Martin, 323.

62. "City Intelligence," *New York Times,* 7 December 1858, 4.

63. Jon A. Peterson, "The Impact of Sanitary Reform upon American Urban Planning, 1840–1890," *Journal of Social History* 13.1 (Autumn 1979): 83, 86.

64. Martin V. Melosi, *The Sanitary City: Environmental Services in Urban America from Colonial Times to the Present* (Pittsburgh: University of Pittsburgh Press, 2008), 14–15, 31.

65. Egbert L. Viele, *The Topography and Hydrology of New York* (New York: Robert Craighead, 1865), 4–8; Melosi, 29–30, 31.

66. Peterson, 84, 86.

67. Melosi, 43–45.

68. Ellis L. Armstrong, *History of Public Works in the United States, 1776–1976* (Chicago: American Public Works Association, 1976), 434.

69. Peterson, 90–91; Melosi, 99–102.

70. Hoy, 68. By 1879, Waring had made a name for himself as a champion of sanitation and sewer design.

71. Peterson, 92.

72. For a complete listing of Olmsted's projects, see the website for the documentary film *Olmsted and America's Urban Parks,* accessed 26 August 2011 at http://www.theolmstedlegacy.com/parks/. According to the website's count, Olmsted designed 140 parks, parkways, recreation areas, and scenic reservations; 20 subdivisions and suburban communities; 34 college and school campuses; 6 institutional grounds; 19 grounds of public buildings; and 5 cemeteries and memorials.

73. Peter Salwen, *Upper West Side Story: A History and Guide* (New York: Abbeville, 1989), 59; Hoy, 62–67.

74. Egbert L. Viele, "Report," preface. The six committee members were Viele, Charles H. Haswell of New York, Dr. E. M. Snow of Rhode Island, Dr. Henry Guernset of New York, Henry Irwin of Virginia, and Otis Clapp of Massachusetts.

75. Egbert L. Viele, "Report," 6.

76. Peterson, 84, 91. One of the first scientists to isolate and discover the nexus between microorganisms and disease was a German, Robert Koch. In 1883, while in Egypt, he found that the

Vibrio comma, "a motile, comma-shaped bacterium," caused cholera. See Charles E. Rosenberg, *The Cholera Years: The United States in 1832, 1849, and 1866* (Chicago: University of Chicago Press, 1987), 3. Also see Melosi, 42, 77.

77. *Real Estate Record and Builders' Guide* 43 (15 June 1889): 832; "Dwelling-house Sanitation," *Real Estate Record and Builders' Guide* 35 (27 June 1885): 729; Melosi, 67.

78. Egbert L. Viele, "Report," 34–36. Baron Haussmann of Louis Napoleon's Paris influenced the tastes and desires of New York's urban developers. When the apartment house became vogue in the 1870s, New Yorkers mostly resisted the "French flats" behind Haussmann's tree-lined boulevards. But the city's "most francophiliac architect," Rutherford Stuyvesant, hired Richard Morris Hunt in 1869 to build a stretch of four row houses on 18th Street between Irving Place and Third Avenue, creating the first French-style apartment flats in the city. They proved to be popular and were quickly filled with renters. Paris remained a "glittering capital" that "spellbound" the elite of New York City in the decade after the Civil War. See Burrows and Wallace, 971, 1002.

79. "The New York Sanitary Association," *New York Times,* 10 June 1859, 4.

80. Egbert L. Viele, *The Topography and Hydrology of New York,* and fold map (USMA Special Collections).

81. Cohen and Augustyn, 136–37.

82. Egbert L. Viele, *The Topography and Hydrology of New York,* 4.

83. Ron Chernow, "The Silent Springs of Manhattan," *New York Times,* 6 March 1977, 1.

84. Steven Kurtz, "City Lore; When There Was Water, Water Everywhere," *New York Times,* 11 June 2006, 4; Cohen and Augustyn, 136–37; Chase Viele, "A Short Biography of Egbert L. Viele" (manuscript), 35.

85. Egbert L. Viele, *The Topography and Hydrology of New York,* 4, 11.

86. Rosenberg, 4–5.

87. Hoy, 29–59. Hoy provides a useful summary of the Sanitary Commission and the work of women nurses during the Civil War. After the Civil War, more Americans realized the deadly connections between dirt and disease.

88. *Annual Reunion* (1902), 142; Emmons Clark, *History of the Seventh Regiment of New York, 1806–1889,* vol. 2 (New York, 1890), 50; Ezra J. Warner, *Generals in Blue: Lives of the Union Commanders* (Baton Rouge: Louisiana State University Press, 1964), 527.

89. Egbert L. Viele, *Handbook for Active Service; Containing Practical Instructions in Campaign Duties for the Use of Volunteers* (New York: D. Van Nostrand, 1861), 1–8; Warner, 527. According to Jay Henry Mowbray, the Confederates reprinted the Viele manual without Viele's consent, and he confiscated copies found during his time in Norfolk. See Jay Henry Mowbray, ed., *Representative Men of New York: A Record of Their Achievements* (New York: New York Press, 1898), 2:165.

90. Warner, 527; Elise Strother Tuckerman, *The Pendulum Swings* (New York: Vantage, 1962), 29. In the spring of 1861, Viele made presentations on fortifications and sought to form a brigade of volunteers under his leadership to fight the rebellion. See "Lecture on Military Science by Lieut. Viele," *New York Times,* 28 February 1861, 8; and "Capt. Viele," *New York Times,* 15 June 1861, 5.

91. *Annual Reunion* (1902), 143; Clark, *History of the Seventh Regiment of New York, 1806–1889,* vol. 2, 50.

92. Warner, xv–xxv. During the Civil War, both the United States and the Confederacy lacked a large pool of senior officers to lead their respective militaries. Of the 583 Union generals, only 194

were professional soldiers, and 217 were graduates of West Point. More were "political" generals who received their appointments based on their civilian roles and responsibilities before the war. The result was that officers quickly rose to the rank of either brigadier general or major-general because they were "best-placed instead of best-qualified."

93. Burrows and Wallace, 885; James M. McPherson, *Tried by War: Abraham Lincoln as Commander in Chief* (New York: Penguin, 2008), 152. Supporters of a separate Northwest Confederacy hoped to reopen the Mississippi River to farm goods produced in the Midwest. Gen. John McClernand was a "political" general with limited military experience. He had been a member of Congress as a Kentucky representative who moderately supported the South's positions on slavery and disunion. Lincoln used the appointment of McClernand as a way to keep Democrats from southern Illinois in the Union. See Warner, 293.

94. Edward K. Spann, *Gotham at War: New York City, 1860–1865* (Wilmington, DE: Scholarly Resources, 2002), 89–90; Mushkat, 138–57.

95. Spann, *Gotham at War,* 83–85.

96. Egbert L. Viele, "A Trip with Lincoln, Chase, and Stanton," *Scribner's Monthly* 16.6 (October 1878): 813; Egbert L. Viele, "Lincoln as Story Teller," in *Abraham Lincoln: Tributes from His Associates, Reminiscences of Soldiers, Statesmen, and Citizens* (New York: Thomas Y. Crowell, 1895), 116–24.

97. Egbert L. Viele, "A Trip with Lincoln, Chase, and Stanton," 813.

98. Egbert L. Viele, "Lincoln as Story Teller," 118.

99. Chase Viele, "Biography of Egbert L. Viele," 26; Richard Henry Savage, "General Egbert Ludovicus Viele," reprinted in Kathlyne Knickerbocker Viele, *Viele, 1659–1909: Two Hundred and Fifty Years with a Dutch Family of New York* (New York: Tobias A. Wright, 1909), 127–30.

100. Savage, 123–24; Ervin L. Jordan, *Black Confederates and Afro-Yankees in Civil War Virginia* (Charlottesville: University Press of Virginia, 1995), 256.

101. Kathlyne Knickerbocker Viele, *Viele, 1659–1909,* 118. Egbert L. Viele Jr. was the seventh child of General Viele and Teresa Griffin Viele. After the Vieles divorced in 1872, Egbert Jr. went to Paris with his mother and changed his name to Francis Viele-Griffin. As Francis Viele-Griffin, he became known as a French poet of the symbolist movement.

102. Reinhard Clifford Kuhn, *The Return to Reality: A Study of Francis Viele-Griffin* (Paris: Librairie Minard, 1962), 9; Jordan, 178–80.

103. "General Dix's Department," *New York Times,* 26 February 1863, 8; "From Fortress Monroe," *New York Herald,* 23 February 1863, 5.

104. "Washington's Birthday in Norfolk," *New York Herald,* 26 February 1863, 8.

105. Warner, 527.

106. Letter of Resignation, Egbert L. Viele to the President of the U.S., 16 October 1863 (USMA Special Collections, NA RG 94).

107. Advertisement, *New York Herald,* 30 January 1864, 9.

108. Burrows and Wallace, 896. Additionally, Tammany came up with a plan to pay the $300 substitution fee for the poorer, working-class constituents of New York. They drafted an ordinance "to relieve the City of New York from unequal operation of conscription and to encourage volunteering" and then "appropriated $2,500,000 to pay the three-hundred-dollar exemption for every New York conscript." The city sold conscription exemption bonds at 7 percent interest, which

could be redeemed by 1880. Iver Bernstein notes that "[the] nature and extent of the proposed expenditure were without precedent." Tammany's strategy upset conservatives because the great giveaway would change the political control of the city. It also reflected the conciliatory nature that elites of both parties applied toward resolving the crisis of the draft riots. See Bernstein, 52.

109. Irving Katz, *August Belmont: A Political Biography* (New York: Columbia University Press, 1968), 124; Burrows and Wallace, 899.

110. Ernest A. McKay, *The Civil War and New York City* (Syracuse, NY: Syracuse University Press, 1990), 235–45.

111. Kuhn, 19.

112. *Real Estate Record and Builders' Guide* 2 (3 October 1868).

113. David McCullough, *The Greater Journey: Americans in Paris* (New York: Simon and Schuster, 2011), 206–207.

114. Burrows and Wallace, 923; David M. Scobey, *Empire City: The Making and Meaning of the New York City Landscape* (Philadelphia: Temple University Press, 2002), 190; *Real Estate Record and Builders' Guide* 2 (3 October 1868).

115. "General City News," *New York Times,* 4 October 1866, 8.

116. "Local Intelligence," *New York Times,* 22 October 1868, 2.

117. Burrows and Wallace, 929.

118. Salwen, *Upper West Side Story,* 55.

119. *West Side Association Minutes* (New-York Historical Society: MSS, 1879–1885, 1893), 22 November 1879. Hereafter abbreviated to *WSAM.*

120. Burrows and Wallace, 924, 925.

121. Salwen, *Upper West Side Story,* 60, 64, 72.

122. *WSAM,* April–June 1879, 4–40, 48–50. According to the record, "Messrs. Dwight H. Olmsted, John W. Pirsson, Cyrus Clark, George H. Peck, Wm. H. Scott, Howard W. Coates, Simeon E. Church, N. P. Bailey, F. H. Cossitt, Genl. Egbert L. Viele and others" were all present at the first meeting.

123. "The Building Movement," *New York Times,* 21 December 1879, 6; and Rosenzweig & Blackmar, 291.

124. "West Side Improvements," *New York Times,* 6 March 1881, 2.

125. "Division of the Offices, Mayor Edson's Nominations as Dictated by Kelly," *New York Times,* 10 January 1883, 8; and *Register of Graduates and Former Cadets of the United States Military Academy* (2000).

126. Quoted in Peter Salwen, "Past Tense: Soldier of Misfortune," *New York Alive* (Jan.–Feb. 1990): 16.

127. "West Side Improvements," *New York Times,* 6 March 1881, 2.

128. Salwen, *Upper West Side Story,* 74–75.

129. Ibid., 72. Today, there is a twelve-story high-rise apartment building (155 Riverside Drive) on the site. Clearly, the forces of New York's real estate market overwhelmed the "mansion neighborhood" that Viele and his West Side peers had envisioned for the West End plateau. Viele's house was demolished shortly after his death, and a series of high rises were constructed in its place. One of the families to live in the building at 155 Riverside Drive was J. Robert Oppenheimer, "father of the atomic bomb," and his parents, who had an eleventh-story apartment overlooking

the Hudson River. See William J. Broad, "Why They Called It the Manhattan Project," *New York Times,* 30 October 2007, 1.

130. Egbert L. Viele, "A Glimpse of Nature from My Veranda," *Harper's New Monthly Magazine* 57 (August 1878): 405.

131. Salwen, *Upper West Side Story,* 72, 74–75.

132. "Ten Thousand Squatters," *New York Times,* 20 April 1880, 8.

133. "The Site for the World's Fair," *New York Times,* 6 November 1880, 8; "Values Then and Now," *Real Estate Record and Builders' Guide* 26 (11 December 1880): 1084–85; "Asking for Improvements," *New York Times,* 14 June 1885, 2; "West Side Property Holders," *New York Times,* 12 July 1885, 12; *WSAM,* 13–50.

134. Charles W. Cheape, *Moving the Masses: Urban Public Transit in New York, Boston, and Philadelphia, 1880–1912* (Cambridge, MA: Harvard University Press, 1980), 28.

135. Domenic Vitiello, "Monopolizing the Metropolis: Gilded Age Growth Machines and Power in American Urbanization," *Planning Perspectives* 28.1 (January 2013): 73, 85.

136. Egbert L. Viele, letter dated 16 November 1868, in *Robert A. Chesebrough's System of Locomotion for Elevated Railroads* (New York: C. S. Westcott, 1869), 11; Henry Hall, ed., *America's Successful Men of Affairs: An Encyclopedia of Contemporary Biography* (New York: The New York Tribune, 1895), 1:139. Chesebrough reportedly consumed a spoonful of Vaseline daily, living to the age of ninety-six. See E. Schwager, "From Petroleum Jelly to Riches," *Drug News and Perspectives* 11.2 (1998): 127.

137. Egbert L. Viele, letter dated 16 November 1868, in *Chesebrough's System of Locomotion,* 4–5, 11, 22.

138. Egbert L. Viele, *The Arcade Under-Ground Railway* (New York, 1872), cover. As noted in chapter 4, Adams spent a year at West Point before dropping out. All three were integral members of the ASCE, and McCallum was a Scottish-born engineer who organized Union Army railroads as a major-general during the Civil War. McCallum's son, William B. McCallum, was a member of the West Point class of 1867. See *Annual Reunion* (1889), 97.

139. Egbert L. Viele, *The Arcade Under-Ground Railway,* 1–3.

140. Cheape, 22, 33, 36.

141. "West Side Improvements," *New York Times,* 6 March 1881, 2.

142. *Real Estate Record and Builder's Guide* 28 (26 November 1881).

143. "New Parks Far Up Town," *New York Times,* 19 March 1882, 14; New York City Department of Parks & Recreation website, accessed 20 June 2011 at http://www.nycgovparks.org /sub_about/parks_history/commissioners.html.

144. "The Railroad at the Battery," *New York Times,* 11 February 1883, 9. In a hearing on the construction of a rail station at Battery Park, Viele indicated that he thought the station needed to be built, but if the rail line had to cross through Battery Park, it should at least be elevated and "made an ornamental structure." Viele was clearly not in the camp of landscape aesthetes like Olmsted.

145. Herman Knickerbocker Viele, "General Egbert L. Viele," in *The New York Genealogical and Biological Record* (New York: New York Genealogical and Biographical Society, 1903), 34.1 (January 1903): 4, 6.

146. Tuckerman, 26.

1. David McCullough, *The Great Bridge: The Epic Story of the Building of the Brooklyn Bridge* (New York: Simon and Schuster, 1982), 104, 111–12.

2. John Forster, *The Life of Charles Dickens,* vol. 3, *1852–1879* (Philadelphia: J. B. Lippincott, 1874), 410.

3. Daniel M. Bluestone, "From Promenade to Park: The Gregarious Origins of Brooklyn's Park Movement," *American Quarterly* 39.4 (Winter 1987): 542; F. L. Olmsted to Calvert Vaux, letter, 12 March 1865, in *The Papers of Frederick Law Olmsted,* vol. 5, *The California Frontier, 1863–1865,* ed. Victoria Post Ranney (Baltimore: Johns Hopkins University Press, 1990), 324; Egbert L. Viele, "Prospect Park, Report, January 15, 1861," in *First Annual Report of the Commissioners of Prospect Park* (Brooklyn, January 28, 1861), 34.

4. Roy Rosenzweig and Elizabeth Blackmar, *The Park and the People: A History of Central Park* (Ithaca, NY: Cornel University Press, 1992), 380–81.

5. Edwin G. Burrows and Mike Wallace, *Gotham: A History of New York City to 1898* (New York: Oxford University Press, 1999), 1231–34.

6. Since the colonial era (1693–1776), there were ferries and bridges connecting Manhattan with the Bronx across Spuyten Duyvil and the Harlem River, but circumnavigation of Manhattan by larger vessels was not possible until the opening of the Harlem Ship Canal in 1895. See Sharon Reier, *The Bridges of New York* (New York: Quadrant, 2000), 66–70.

7. Burrows and Wallace, 1231; George J. Lankevich, *New York City: A Short History* (New York: New York University Press, 2002), 135.

8. Theodore Roosevelt, *New York* (New York: Longmans, Green, 1891), 214.

9. USMA Association of Graduates, *Annual Reunion of the Association of Graduates of the United States Military Academy,* 10 June 1895, 110.

10. Brian C. Melton, *Sherman's Forgotten General: Henry W. Slocum* (Columbia: University of Missouri Press, 2007), 216, 224.

11. Burrows and Wallace, xx–xxi. For insight into the age, see Kristin L. Hoganson, *Consumers' Imperium: The Global Production of American Domesticity, 1865–1920* (Chapel Hill: University of North Carolina Press, 2007), 23, 62–67.

12. McCullough, 547; Melton, 228.

13. "The *General Slocum* an Unlucky Craft," *New York Times,* 16 June 1904, 1.

14. *Annual Reunion* (1894), 80–84.

15. Melton, 206.

16. "Gen. Slocum's Position," *New York Times,* 5 October 1865, 8.

17. Melton, 213–14.

18. Ibid., 219; "Gen. Slocum's Speech," *New York Times,* 5 October 1865, 4.

19. Melton, 225; Charles E. Slocum, *The Life and Services of Major General Henry Warner Slocum* (Toledo, OH: Slocum, 1913), 333.

20. Melton, 10–13, 15–16. Melton lists James B. McPherson, John Bell Hood, William Carlin, Thomas Ruger, Jeb Stuart, and Stephen D. Lee as other notable Civil War figures that were at West Point at the same time as Slocum.

21. Slocum, 10; Philip H. Sheridan, *Personal Memoirs of P. H. Sheridan* (New York: Charles L. Webster, 1888), 1:10.

22. George W. Cullum, *Biographical Register of the Officers and Graduates of the U.S. Military Academy*, 3rd ed. (New York: Houghton, Mifflin, 1891), 2:476.

23. Melton, 21–22; Cullum, 476.

24. Melton, 29.

25. Slocum, 11; Melton, 30.

26. Melton, 246. See *The War of Rebellion: A Compilation of the Official Records of the Union and Confederate Armies*, 128 vols. (Washington, DC: GPO, 1881–1901), Series 1.

27. Melton, 63–64; *The War of Rebellion* 5:236.

28. See James M. McPherson, *Tried by War: Abraham Lincoln as Commander in Chief* (New York: Penguin, 2008), 65–134.

29. Melton, 69.

30. Ibid., 105–108.

31. Ibid., 148; Ezra J. Warner, *Generals in Blue: Lives of the Union Commanders* (Baton Rouge: Louisiana State Press, 1964), 452.

32. Warner, 452; *Annual Reunion* (1894), 83.

33. *Annual Reunion* (1894), 84; Melton, 175.

34. W. T. Sherman to H. W. Slocum, 13 March 1868, in Slocum, 334.

35. George Lankevich, "The Grand Army of the Republic in New York State, 1865–1898" (PhD diss., Columbia University, 1967), 39.

36. Melton, 238–40; "A Nation Is Mourning," *New York Times*, 16 February 1891, 1.

37. *Annual Reunion* (1894), 78.

38. "Funeral of Gen. H. W. Slocum," *New York Times*, 18 April 1894, 9. Slocum's pallbearers included West Pointers George S. Greene and Fitz John Porter; Medal of Honor recipients Benjamin F. Tracy, Dan E. Sickles, Dan Butterfield, and Horatio C. King; Justices Edgar Cullen and Calvin Pratt; and former New York mayor Abram S. Hewitt. Sickles was renowned for lurid scandals and a lack of scruples. He even got away with shooting Francis Scott Key's son over a woman. See Warner, 378–80, 446–47.

39. Melton, 227; McCullough, 83; "John A. Roebling," *Brooklyn Daily Eagle*, 26 July 1869, 2, where the reporter noted that "no officer came out of the late war with a brighter record than General Slocum."

40. Melton, 227–28.

41. Ibid., 228; McCullough, 138.

42. Melton, 228, 231; "Local Politic," *Brooklyn Daily Eagle*, 5 November 1870, 2.

43. McCullough, 90.

44. *Brooklyn Daily Eagle*, 26 July 1869, 2.

45. Louis H. Sullivan, *The Autobiography of an Idea* (New York: American Institute of Architects, 1924), 6, 247.

46. McCullough, 49–50.

47. "Mr. Roebling's Death," *Brooklyn Daily Eagle*, 22 July 1869, 1.

48. McCullough, 36, 468.

49. Ibid., 35.

50. Ibid.; Hampton L. Carson, "Andrew Athinson Humphreys, Brigadier-General U.S. Army, Brevet Major-General, Chief of Engineers," *Proceedings of the American Philosophical Society* 22 (January 1885): 52, 67.

51. McCullough, 35.

52. *Annual Reunion* (1900), 32; *Annual Reunion* (1895), 106; Cullum, *Biographical Register of the Officers and Graduates of the United States Military Academy*, 3rd ed. (New York: Houghton Mifflin, 1891), 2:869.

53. McCullough, 23.

54. "John A. Roebling," *Brooklyn Daily Eagle*, 26 July 1869, 2; McCullough, 462–64; Ken Burns, *Brooklyn Bridge*, PBS documentary film (1982), accessed 31 July 2011, http://www.pbs.org/kenburns /brooklynbridge/. This documentary and McCullough's book are two of the most popular sources for celebrating the role of Emily Warren Roebling in the building of the Brooklyn Bridge. Also see Marilyn E. Weigold, *Silent Builder: Emily Warren Roebling and the Brooklyn Bridge* (New York: Associated Faculty Press, 1984).

55. McCullough, 153, 160, 163.

56. *Annual Reunion* (1883), 28, 33.

57. McCullough, 462. G. K. Warren graduated second in the West Point class of 1850.

58. McCullough, 457; *Annual Reunion* (1883), 28.

59. Ibid. Gouverneur Kemble was no stranger to West Point, routinely hosting Military Academy faculty and officers at his river-side mansion for Saturday evening dinner parties, especially during the Thayer years. See Theodore Crackel, *West Point: A Bicentennial History* (Lawrence: University Press of Kansas, 2002), 94–95.

60. McCullough, 458; *Register of Graduates* (2000 ed.), 4–31.

61. Weigold, 4.

62. George Parsons Lathrop and Rose Hawthorne Lathrop, *Story of Courage: Annals of the Georgetown Convent of the Visitation of the Blessed Virgin Mary* (Cambridge, MA: Riverside, 1894), 366–69. Many of the girls who attended the Visitation Academy married prominent figures of the nineteenth century. Among them were the wives of Union generals Sherman (Ellen Ewing) and Bache (Minnie Meade), and Confederate generals Joseph E. Johnston (Lydia M. S. McLain), and Pierre Gustave Toutant Beauregard (Caroline Deslonde). Stephen A. Douglas also married a Visitation graduate (Adelaide Cutts). Also attending the school was President Buchanan's niece Harriet Lane Johnson. Clearly, Visitation Academy was a place where Warren hoped his sister would receive the best preparation to be a woman of substantial status for the time.

63. *Annual Reunion* (1883), 31–36.

64. McCullough, 160; *Annual Reunion* (1883), 32.

65. McCullough, 161, 459.

66. Ibid., 462–63; Weigold, 35–36.

67. McCullough, 472.

68. "The Woman's Law Class," *New York Times*, 31 March 1899, 1. This was the ninth class that included women to complete the law program. Emily's speech was titled "The Wife's Disabilities."

69. McCullough, 343, 474–75.

70. "Obituary: Col. William H. Paine," *Street Railway Journal* 7:1 (New York, February 1891), 95.

71. G. K. Warren to Secretary of War Stanton, Subj.: Recommending William H. Paine for promotion, 28 February 1865, in William H. Paine Papers, MS 475, New-York Historical Society.

72. "Obituary: Col. William H. Paine," 95.

73. Emily Warren Roebling to William H. Paine, 31 January 1872, in Paine Papers, MS 475, New-York Historical Society.

74. Emily Warren Roebling to Mrs. William H. Paine (Catherine), 27 December 1877, in Paine Papers.

75. Emily Warren Roebling to Mrs. William H. Paine (Catherine), 27 February 1880, in Paine Papers; "Current Events," *Brooklyn Daily Eagle,* 26 February 1880, 2.

76. "The East River Bridge. Successful Launch of the New York Caisson," *Brooklyn Daily Eagle,* 9 May 1872, 4.

77. McCullough, 523.

78. Ibid., 466; "Steel for the Bridge Cables," *New York Times,* 5 May 1879, 2.

79. Melton, 230.

80. Ibid., 230–31; McCullough, 466; "Nothing to Show Bribery," *New York Times,* 6 May 1879, 10.

81. McCullough, 467–68; "Nothing to Show Bribery," *New York Times,* 6 May 1879, 10; "A Majority for Roebling," *New York Times,* 12 September 1882, 8.

82. Weigold, 139–40. Emily died 28 October 1903, leaving an estate valued at $475,000. Washington Roebling lived to the age of eighty-nine, dying in 1926.

83. Ibid., 145; "Brooklyn Honors Bridge's Builders: Plaque to Roebling Family Is Unveiled," *New York Times,* 25 May 1953, 26.

84. Weigold, 145; Burns, *Brooklyn Bridge* film.

85. Photo, West Tower of the Brooklyn Bridge, 24 August 2010, personal collection. The plaques name sixty trustees and eight engineers, but make no mention of Emily Roebling. Slocum's name is fourth on the list.

86. "Not to Go: Mr. Roebling Still in Charge of the Bridge," *Brooklyn Daily Eagle,* 12 September 1882, 2.

87. McCullough, 499, 503–504.

88. Weigold, 45–46; "A Majority for Roebling," *New York Times,* 12 September 1882, 8.

89. McLaughlin, "The 'Boss' in Reply to General Slocum," *Brooklyn Daily Eagle,* 26 October 1875, 4.

90. Melton, 232.

91. "Brooklyn Politics," *Brooklyn Daily Eagle,* 2 January 1898, 52–53.

92. "City Works: What General Slocum Knows about Them," *Brooklyn Daily Eagle,* 8 February 1876, 2; Melton, 232–33.

93. "Heads Off: Reorganizing the Department of City Works," *Brooklyn Daily Eagle,* 9 July 1877, 4; "Municipal: 'Civil Service' with a Vengeance," *Brooklyn Daily Eagle,* 19 July 1877, 4; Melton, 233.

94. "Slocum: The General Disgusted with His Associate Commissioners," *Brooklyn Daily Eagle,* 3 January 1878, 4.

95. McCullough, 504; Raymond A. Mohl, *The New City: Urban America in the Industrial Age, 1860–1920* (Wheeling, IL: Harlan Davidson, 1985), 83–90, 100–101; Lankevich, *New York City: A Short History,* 134.

96. Otto Eisenschiml, *The Celebrated Case of Fitz John Porter: An American Dreyfus Affair* (New York: Bobbs-Merrill, 1950), 288, 290. The Porter case is discussed in greater detail in chapter 7.

97. Melton, 236, 244.

98. "General Slocum's Warning," *Brooklyn Daily Eagle*, 2 July 1890, 4.

99. Samuel E. Tillman, "The Academic History of the Military Academy, 1802–1902," in *The Centennial of the United States Military Academy at West Point, New York, 1802–1902* (Washington, DC: GPO, 1904), 1:353. Jacob Bailey was a professor of chemistry, mineralogy, and geology at West Point.

100. George Templeton Strong, *The Diary of George Templeton Strong*, vol. 2, *The Turbulent Fifties, 1850–1859*, ed. Allan Nevins and Milton Halsey Thomas (New York: Macmillan, 1952), 18, 101.

101. Stephen W. Sears, *George B. McClellan: The Young Napoleon* (New York: Ticknor and Fields, 1988), 13.

102. Ibid., 38, 43.

103. *Annual Reunion* (1886), 57. The Delafield Commission consisted of Majors Richard Delafield and Alfred Mordecai, and Captain McClellan. McClellan's report on the Crimean War further enhanced his reputation as a young military genius. See Matthew Moten, *The Delafield Commission and the American Military Profession* (College Station: Texas A&M University Press, 2000), 73–107.

104. Sears, 63; *Annual Reunion* (1886), 58.

105. Sears, 53, 57, 64, 67.

106. Ibid., 58–60; McPherson, *Tried by War*, 47, 68.

107. McPherson, *Tried by War*, 91, 113, 126, 141.

108. "The Peace Party: The Wood-Vallandigham Faction in Convention," *New York Times*, 4 June 1863, 1.

109. Sven Beckert, *The Monied Metropolis: New York City and the Consolidation of the American Bourgeoisie, 1850–1896* (New York: Cambridge University Press, 2001), 128–29; Irving Katz, *August Belmont: A Political Biography* (New York: Columbia University Press, 1968), 122.

110. Bernstein, 51.

111. Beckert, 133.

112. "Interesting Ceremony at West Point," *New York Times*, 15 June 1864, 5.

113. Sears, 363, 374.

114. Beckert, 134.

115. Katz, 148. Lincoln's victory in the election of 1864 was not just a result of the Democrats' disjointed campaign. Other events factored into the election outcome, including Sherman's victory at Atlanta and ballots cast by Union soldiers in the field. See James McPherson, *Battle Cry of Freedom: The Civil War Era* (New York: Oxford University Press, 1988), 803–805.

116. Sears, 387, 388, 390–92. According to Sears, "Moltke confided details of the recent Austro-Prussian War and flattered him with the observation that his Peninsula campaign of 1862 would surely have ended the Civil War had he not been 'shamefully deserted' by the Lincoln government. (Since Moltke's opinion was no doubt derived from a reading of McClellan's Report, he could hardly have reached any other conclusion.)"

117. McClellan earned $7,849 for the 1868 tax year. Andrew J. Moulder to George B. McClel-

lan, 12 November 1868; and "Receipt for Special Tax," 24 May 1869, in George Brinton McClellan Papers, Library of Congress, microfiche reel 37.

118. D. L. Simpson to George B. McClellan, 14 November 1868, McClellan Papers, reel 37.

119. Sears, 392.

120. Ann L. Buttenwieser, *Manhattan Water-Bound: Manhattan's Waterfront from the Seventeenth Century to the Present*, 2nd ed. (Syracuse, NY: Syracuse University Press, 1999), 65–66.

121. Burrows and Wallace, 950.

122. "Department of Docks," *New York Times,* 19 July 1870, 3.

123. "Improved Piers—Gen. McClellan's Plan," *New York Times,* 30 November 1870, 2.

124. "The Plan of the Dock Commissioners," *New York Times,* 1 June 1871, 4.

125. Burrows and Wallace, 950.

126. Buttenwieser, 77–78.

127. Oakey Hall to George B. McClellan, 16 September 1871, in George Brinton McClellan Papers, Library of Congress, microfiche reel 37.

128. William G. Hunt to George B. McClellan, 18 September 1871; William Averell to George B. McClellan, 19 September 1871; William W. Burns to George B. McClellan, 21 September 1871, all in George Brinton McClellan Papers, Library of Congress, microfiche reel 37.

129. James McHenry to George B. McClellan, 4 May 1872, in George Brinton McClellan Papers, Library of Congress, microfiche reel 37.

130. Sears, 392–93, 397.

131. Ibid., 406. McClellan's son, George Brinton McClellan Jr., was a congressman for New York and also mayor of New York City from 1904 to 1909.

132. "United! Brooklyn and New York by the Great Bridge," *Brooklyn Daily Eagle,* 24 May 1883, 1.

133. McCullough, *The Great Bridge,* 528–37.

134. Burrows and Wallace, 1228; "Glorification! The Two Cities Celebrate the Work That Makes Them One," *Brooklyn Daily Eagle,* 24 May 1883, 12.

135. Reier, 34, 154–55.

136. U.S. Army Corps of Engineers, New York District, "The Conquest of Hell Gate," accessed 2 April 2008 at http://www.nan.usace.army.mil/whoweare/hellgate.pdf, 2–3.

137. *War Department Annual Reports, 1916,* vol. 2, *Report of the Chief of Engineers* (Washington, DC: GPO, 1916), 215.

138. U.S. Army Corps of Engineers, *Analytical and Topical Index to the Reports of the Chief of Engineers and the Officers of the Corps of Engineers, United States Army, upon Works and Surveys for River and Harbor Improvement, 1866–1879* (Washington, DC: GPO, 1881), 194.

139. Marion J. Klawonn, *Cradle of the Corps: A History of the New York District U.S. Army Corps of Engineers, 1775–1975* (New York: U.S. Army Engineer Corps, 1977), 149.

140. *Annual Reunion* (1895), 109; "Fragments of Flood Rock," *New York Times,* 11 October 1885, 1; William Kornblum, *At Sea in the City: New York from the Water's Edge* (Chapel Hill, NC: Algonquin, 2002), 183–204.

141. *Annual Reunion* (1895), 105–106.

142. Ibid., 98–99, 105–106, 110.

143. "Our Harbor Entrances," *New York Times,* 14 July 1869, 8.

144. "Report of the Chief of Engineers," in *Report of the Secretary of War to the 41st Congress,* Vol. 2 (Washington, DC: GPO, 1869), 393–94.

145. *Annual Reunion* (1895), 107, 108.

146. "Harlem Canal Parades," *New York Times,* 17 June 1895, 1.

147. Goldman, 154; Thomas Kessner, *New York City and the Men behind America's Rise to Economic Dominance, 1860–1900* (New York: Simon and Schuster, 2003), 153.

148. Burrows and Wallace, 1224.

149. Kessner, 319–20.

150. Burrows and Wallace, 1050.

151. Kessner, 325–26; Burrows and Wallace, 1230–33.

152. Roosevelt, 214.

153. Strong, *The Diary of George Templeton Strong,* vol. 4, *Post-War Years, 1865–1875,* ed. Nevins and Thomas, 374.

154. For an interesting take on these statues, see Federal Writers' Project, *New York City Guide* (New York: Random House, 1939), 229, 288. For Slocum's statue, see New York City Department of Parks & Recreation website, accessed 3 August 2011 at http://www.nycgovparks.org/parks/grand armyplaza/highlights/11844. Grant's Tomb was built on the Upper West Side in 1897. In Central Park's Grand Army Plaza stands a bronze statue of General Sherman, erected in 1903, and another of Gen. G. K. Warren, which was dedicated in 1896.

155. The population of the United States rose from 23 million to 76 million between 1850 and 1900. In New York, the population tripled as well, going from 515,000 in 1850 to 1.5 million in 1890. U.S. Census website, accessed 3 August 2011 at http://factfinder.census.gov/home/saff/main .html?_lang=en.

156. Lankevich, *New York City: A Short History,* 138–39, 142–43.

157. See James R. Endler, *Other Leaders, Other Heroes: West Point's Legacy to America beyond the Field of Battle* (Westport, CT: Praeger, 1998), 69–72, 122–23. Francis Vinton Greene, class of 1870 and son of George S. Greene, was head of the Barber Asphalt Paving Company and the New York police commissioner in 1903 under Mayor Seth Low. Others achieved success in business like Horace Porter, class of 1860, and Eugene Griffin, class of 1875. Porter was chairman of the executive committee of the city's elevated railway system, and Griffin was General Electric's first vice-president and general manager in 1892.

158. Ibid., 144–46; Kenneth T. Jackson, ed., *The Encyclopedia of New York City* (New Haven, CT: Yale University Press, 1995), 704.

7. REDEMPTION IN POSTBELLUM GOTHAM

1. Otto Eisenschiml, *The Celebrated Case of Fitz John Porter: An American Dreyfus Affair* (New York: Bobbs-Merrill, 1950), 304; Curt Anders, *Injustice on Trial: Second Bull Run, General Fitz John Porter's Court-Martial and the Schofield Board Investigation That Restored His Good Name* (Zionsville, IN: Guild Press, 2002), 415.

2. Ezra J. Warner, *Generals in Blue: Lives of* the *Union Commanders* (Baton Rouge: Louisiana State Press, 1964), 379; James McPherson, *Battle Cry of Freedom: The Civil War Era* (New York: Oxford University Press, 1988), 58–531, 533; Anders, 407–408.

3. USMA Association of Graduates, *Annual Reunion of the Association of Graduates of the United States Military Academy*, 1901, 206; Eisenschiml, 177.

4. Eisenschiml, 297–98; Albert E. Costello, *Our Police Protectors: History of the New York Police from the Earliest Period to the Present Time* (New York: Chas. F. Roper, 1885), 458.

5. Quoted in Eisenschiml, 298; "General Porter at Police Headquarters," *Evening Post*, October 1884, 1.

6. Allan Nevins, *The Evening Post: A Century of Journalism* (New York: Boni and Liveright, 1922), 458, 477. The term "New York's finest" dates back to 1874 when Mayor William Havemeyer first used the phrase to mean the New York Police Department. See James F. Richardson, *The New York Police: Colonial Times to 1901* (New York: Oxford University Press, 1970), 288.

7. See Ernest S. Griffith, *A History of American City Government: The Conspicuous Failure, 1870–1900* (Washington, DC: University Press of America, 1983), 73–75; and David Quigley, *Second Founding: New York City, Reconstruction, and the Making of American Democracy* (New York: Hill and Wang, 2004), 111–13.

8. Raymond A. Mohl, *The New City: Urban America in the Industrial Age, 1860–1920* (Wheeling, IL: Harlan Davidson, 1985), 84; Edward K. Spann, *The New Metropolis: New York City, 1840–1857* (New York: Columbia University Press, 1981), 64–65.

9. Griffith, *A History of American City Government*, 4.

10. Jon C. Teaford, *The Unheralded Triumph: City Government in America, 1870–1900* (Baltimore: Johns Hopkins University Press, 1984), 170–73; also see chapters 5 and 6 herein.

11. Ibid., 133.

12. "No Title," *New York Times*, 2 March 1875, 6; "A New Official," *New York Times*, 3 March 1875, 1; Costello, 458.

13. Kenneth T. Jackson, ed., *The Encyclopedia of New York City* (New Haven, CT: Yale University Press, 1995), 736. For mayoral elections held between 1863 and 1894, the Democrats claimed 13, Republicans 3, and the Independents 1. Of the 13 Democratic mayors, 3 were supported directly by Tammany Hall.

14. Edwin G. Burrows and Mike Wallace, *Gotham: A History of New York City to 1898* (New York: Oxford University Press, 1999), 907; Quigley, 64–65.

15. David C. Hammack, *Power and Society: Greater New York at the Turn of the Century* (New York: Russell Sage Foundation, 1982), 110–11, 114. They were called Swallowtails after the style of frock coats they wore.

16. Burrows and Wallace, 1104, 1108.

17. Burrows and Wallace, 908; Thomas Kessner, *Capital City: New York City and the Men behind America's Rise to Economic Dominance, 1860–1900* (New York: Simon and Schuster, 2003), 47; Richard White, *Railroaded: The Transcontinentals and the Making of Modern America* (New York: Norton, 2011), 14, 31–33.

18. Alfred D. Chandler Jr., *The Visible Hand: The Managerial Revolution in American Business* (Cambridge, MA: Belknap Press of Harvard University Press, 1977), 95; Alan Trachtenberg, *The Incorporation of America: Culture and Society in the Gilded Age* (New York: Hill and Wang, 1982), 58.

19. James McHenry to George B. McClellan, 4 May 1872, George Brinton McClellan Papers, Library of Congress, microfiche reel 37.

20. Eisenschiml, 211. See chapter 4 for a description of George S. Greene's civilian career.

21. Captain William V. Judson, "The Services of Graduates as Explorers, Builders of Railways, Canals, Bridges, Light-Houses, Harbors, and the Like," in *The Centennial of the United States Military Academy at West Point, New York* (Washington, DC: GPO, 1904), 1:846–48.

22. Kessner, 102. Kessner notes that by 1869, over $3 billion in securities was being traded per year in New York. Also see William Cronon, *Nature's Metropolis: Chicago and the Great West* (New York: Norton, 1991), 76–89.

23. Cronon, 96–98; Glenn Porter, *The Rise of Big Business, 1860–1920,* 2nd ed. (Wheeling, IL: Harlan Davidson, 1992), 33.

24. *Annual Reunion* (1931), 82. For example, James H. Wilson was president of the New England and New York Railroad, 1883–1886.

25. Spann, *The New Metropolis,* 63.

26. Burrows and Wallace, 928.

27. *Annual Reunion* (1884), 113–16; *Annual Reunion* (1897), 14–16; "The Whaling Business: Street-Commissioner's Appointments," *New York Times,* 1 May 1858, 8.

28. George J. Lankevich, *New York City: A Short History* (New York: New York University Press, 2002), 100.

29. Leonne M. Hudson, *The Odyssey of a Southerner: The Life and Times of Gustavus Woodson Smith* (Macon, GA: Mercer University Press, 1998), 72.

30. Ibid., 49. Tiemann was affiliated with Tammany, but ran as an independent candidate in the mayoral election of 1857. Woods ran as Tammany's Democratic candidate. See "Daniel F. Tiemann Dead," *New York Times,* 30 June 1899, 7.

31. Anthony Gronowicz, *Race and Class Politics in New York City before the Civil War* (Boston: Northeastern University Press, 1998), 149–50.

32. Fernando Wood's two brothers, Henry and Benjamin, benefited from Wood's influential power in New York. Benjamin Wood was a congressman who wrote an acclaimed "Peace Democratic novel" during the Civil War. After G. W. Smith left New York in 1861, Mayor Wood put Henry Wood in the Streets Department. See Jerome Mushkat, *Fernando Wood: A Political Biography* (Kent, OH: Kent State University Press, 1990), 125.

33. Hudson, 70–71; "The Street Department," *New York Times,* 25 October 1860, 4; Mushkat, 110–11.

34. Hudson, 71.

35. Ibid., 69, 70; Mushkat, 111–12; Tyler Ambinder, "Fernando Wood and New York City's Secession from the Union: A Political Appraisal," *New York History* 68 (January 1987): 92.

36. Hudson, 73–74.

37. "City and County Affairs," *New York Times,* 21 September 1861, 8.

38. Hudson, 75.

39. "Street Commissioner's Office," *New York Times,* 20 September 1861, 4; *Annual Reunion* (1884), 116–17.

40. Hudson, 72; Mushkat, 116–18.

41. "The Late Street Commissioner, Gustavus W. Smith," *New York Times,* 18 October 1861, 5.

42. *Annual Reunion* (1884), 118.

43. Quoted in Stephen W. Sears, *George B. McClellan: The Young Napoleon* (New York: Ticknor and Fields, 1988), 66.

44. "Gen. Mansfield Lovell," *New York Times,* 8 April 1862, 5.

45. "At Fault in His Classics: Maj.-Gen. Mansfield Lovell," *New York Times,* 24 May 1862, 4.

46. *Annual Reunion* (1884), 126; "Gen. Mansfield Lovell," *New York Times,* 2 June 1884, 5; Cullum, *Biographical Register* (1891), 2:122.

47. "Aldermanic Measures," *New York Times,* 9 July 1875, 2.

48. "Not Fit to Be Made," *New York Times,* 16 November 1878, 4.

49. "Gen. Mansfield Lovell," 5.

50. *Annual Reunion* (1884), 117.

51. Stephen E. Ambrose, *Duty, Honor, Country: A History of West Point* (Baltimore: Johns Hopkins University Press, 1999), 188–89. Specifically, in accordance with the wishes of George W. Cullum, superintendent after the war, Cullum Memorial Hall, built in his honor, was to have no mention of graduates who defected to the South. Also, Battle Monument on Trophy Point lists only fallen Union soldiers from the regular Army. Also see Frank J. Walton, "The West Point Centennial: A Time for Healing," in Lance Betros, ed., *West Point: Two Centuries and Beyond* (Abilene, TX: McWhiney Foundation Press, 2004), 209–14.

52. *Annual Reunion* (1897), 20–21; Gustavus W. Smith, *Life Insurance: The Practical Business* (New York: D. Van Nostrand, 1878).

53. Quigley, xiv–xv, 183.

54. Eisenschiml, 210.

55. "Another Groan from a Democrat," *New York Times,* 4 March 1875, 7.

56. George Templeton Strong, *The Diary of George Templeton Strong,* vol. 4, *Post-War Years, 1865–1875,* ed. Allan Nevins and Milton Halsey Thomas (New York: Macmillan, 1952), 553.

57. "The Department of Public Works," *New York Times,* 14 March 1876, 2.

58. "Tammany Leaders Swallowing up the Patronage Expected by the Rank and File," *New York Times,* 4 April 1875, 2.

59. Fitz John Porter to William H. Wickham, 30 April 1875, in Samuel J. Tilden Papers, New York Public Library MSS, Box 20.54.

60. Fitz John Porter to Samuel J. Tilden, 27 May 1875, in Tilden Papers.

61. Eisenschiml, 210.

62. "The Department of Public Works," 1; Burrows and Wallace, 1027–28.

63. "Gen. Porter Rejected," *New York Times,* 14 January 1876, 1.

64. Eisenschiml, 210.

65. *Annual Reunion* (1895), 110.

66. "Gen. Newton Takes the Oath," *New York Times,* 29 August 1886, 3.

67. "Gen. Newton's Career," *New York Times,* 26 August 1886, 1; "The Aqueduct," *New York Times,* 3 May 1888, 8.

68. "City and Suburban News," *New York Times,* 5 October 1886, 3.

69. David Hochfelder, *The Telegraph in America, 1832–1920* (Baltimore: Johns Hopkins University Press, 2012), 47; "W. A. A. Carsey's Convention," *New York Times,* 11 October 1878, 1.

70. Lankevich, *New York City: A Short History*, 111; Lincoln Steffens, "New York: Good Government in Danger," *McClure's Magazine* 22.1 (November 1903): 88–90.

71. Burrows and Wallace, 1108–09.

72. *Annual Reunion* (1895), 110–11.

73. "Gen. John Newton Dead," *New York Times*, 2 May 1895, 3.

74. *Annual Reunion* (1895), 111.

75. Lovell was born in Washington, DC, and Smith came from Georgetown, Kentucky. *Annual Reunion* (1884), 113; Hudson, 4.

76. Costello, 263.

77. James F. Richardson, 164.

78. Ibid., 82–85; Mushkat, 41–42; Strong, *The Diary of George Templeton Strong*, vol. 2, *The Turbulent Fifties, 1850–1859*, ed. Nevins and Thomas, 211–12.

79. Mushkat, 59, 75.

80. Herbert Asbury, *The Gangs of New York: An Informal History of the Underworld* (New York: Alfred A. Knopf, 1928), 100–103.

81. Costello, 142, 152.

82. James F. Richardson, 121, 152–53. Roundsmen were forerunners to the police sergeants. Both had supervisory responsibilities to make sure that the patrolmen were enforcing the laws justly and honestly.

83. Ibid., 154–62; Burrows and Wallace, 928.

84. Costello, 239; Joanne Abel Goldman, *Building New York's Sewers: Developing Mechanisms of Urban Management* (West Lafayette, IN: Purdue University Press, 1997), 150–51.

85. James F. Richardson, 216; Goldman, 150–51.

86. James F. Richardson, 217–18. "Municipal Topics: Arbitrary Dismissal of the Commissioners," *New York Times*, 13 November 1874, 3; "Mayor Havemeyer and His Successor," *New York Times*, 21 November 1874, 2; "Death of the Mayor," *New York Times*, 1 December 1874, 1.

87. "Police Reorganization," *New York Times*, 30 April 1875, 5.

88. "The 'Swallow-Tail' Ascendency," *New York Times*, 30 April 1875, 2.

89. *Annual Reunion* (1903), 146–51; *Official Register of the Office and Cadets of the United States Military Academy* (June 1845), 7; George W. Cullum, *Biographical Register of the Officers and Graduates of the Military Academy at West Point*, vol. 2, no. 1234, 3rd ed. (New York: Houghton Mifflin, 1891), 210.

90. Warner, 463–64.

91. *Annual Reunion* (1903), 198.

92. "Passengers Arrived," *New York Times*, 10 January 1875, 12.

93. "Public Pluck," *New York Times*, 9 August 1875, 4.

94. Lankevich, *New York City: A Short History*, 85.

95. "The Police Commission," *New York Times*, 21 August 1875, 8.

96. James F. Richardson, 219, 221.

97. "Ex-Commissioner Smith," *New York Times*, 3 February 1880, 3.

98. "Gen. Smith Scores a Point," *New York Times*, 11 July 1880, 12.

99. "A Departmental Squabble," *New York Times*, 23 April 1876, 10.

100. James F. Richardson, 220–21.

101. "The Police Raids," *New York Times,* 27 May 1876, 10.

102. James F. Richardson, 180, 193; *Documents of the Senate of the State of New York* (1895), 9:952–53; *Documents of the Senate of the State of New York* (1895), 13:5601–5605. The Lexow Committee was a large-scale state legislature inquiry into two decades of police brutality in the city.

103. James F. Richardson, 209; "The Police Raids," *New York Times,* 27 May 1876, 10.

104. "A Singular Omission," *New York Times,* 23 January 1879, 8.

105. "Police Board Economy," *New York Times,* 25 March 1878, 8; "Gen. Smith Dissatisfied," *New York Times,* 1 December 1880, 8; "A New Police Commissioner," *New York Times,* 12 March 1881, 8.

106. "A New Police Commissioner," *New York Times,* 12 March 1881, 8.

107. William F. Smith did serve briefly as the rapid transit commissioner in the autumn of 1881, but resigned shortly after. "The Mayor and the Police Commissioners," *New York Times,* 2 August 1881, 3.

108. *Annual Reunion* (1903), 199. In 1889, Smith was put on the retired list as a major and received a military pension. Interestingly, he was also receiving federal pay as a civilian civil engineer, and after an inquiry, Smith was allowed to "double dip" as we say today.

109. Eisenschiml, 296; "Fitz John Porter's Relief," *New York Times,* 7 May 1882, 2.

110. "A New Police Commissioner," *New York Times,* 29 October 1884, 8.

111. Eisenschiml, 298; "Mr. Thompson's Successor," New York Times, 7 December 1884, 1.

112. "City and Suburban News," *New York Times,* 31 October 1884, 8.

113. See Fitz John Porter Papers, Library of Congress, MSS 36590 (microfiche reels 1–2, 35).

114. Anders, 415; "The Police Captain's Dinner," *New York Times,* 27 January 1885, 5.

115. "Don't Believe in Fitz John Porter," *New York Times,* 19 February 1886, 1.

116. Jean Edward Smith, *Grant* (New York: Simon and Schuster, 2001), 622, 624; editorial, *New York Times,* 17 January 1885, 4.

117. Editorial, *New York Times,* 19 January 1885, 4.

118. "Gen. Grant's Retirement," *New York Times,* 5 March 1885, 1; Jean Edward Smith, 625.

119. "City and Suburban News," *New York Times,* 13 January 1886, 8.

120. Anders, 415.

121. "Pining to Investigate," *New York Times,* 13 February 1885, 2.

122. "The Police as Ticket Peddlers," *New York Times,* 4 April 1885, 8.

123. "Gen. Porter's Place," *New York Times,* 4 May 1888, 8; Lankevich, *New York City: A Short History,* 130. Hewitt was also a friend and business associate of George B. McClellan. See Sears, 387, 401.

124. Lankevich, *New York City: A Short History,* 129.

125. Eisenschiml, 314.

126. James F. Richardson, 248–67.

127. Sears, 93, 101; "Late News from the South," *New York Times,* 30 June 1862, 1; "The Late Geo. B. McClellan," *New York Times,* 1 May 1863, 8.

128. Eisenschiml, 314.

129. "The Birmingham-Clarke Wedding," *New York Times,* 4 February 1887, 5; "Happy Jeffersonians," *New York Times,* 22 February 1887, 2; "Funeral of Dr. Agnew," *New York Times,* 22 April 1888, 9.

130. Eisenschiml, 314, 318.

131. For example, see "The City Democracy," *New York Times,* 14 March 1875, 1; "Public Pluck," *New York Times,* 9 August 1875, 4; "Departures for Europe," *New York Times,* 10 November 1877, 8; and "A New Police Commissioner," *New York Times,* 29 October 1884, 8.

132. Carl Von Clausewitz, the great Prussian War theorist, explained this characteristic of military thinking, certainly for the nineteenth century, when he wrote: "If we ask what sort of mind is likeliest to display the qualities of military genius, experience and observation will both tell us that it is the inquiring rather than the creative mind, the comprehensive rather than the specialized approach, the calm rather than the excitable head to which in war we would choose to entrust the fate of our brothers and children, and the safety of and honor of our country." Carl Von Clausewitz, *On War,* ed. and trans. Michael Howard and Peter Paret (Princeton, NJ: Princeton University Press, 1984), 112.

133. For the significance of discipline in the cadet education, see Board of Visitors, *Report of the Board of Visitors* (1826), 5.

134. James F. Richardson, 264–65; Arthur S. Link and Richard L. McCormick, *Progressivism* (Wheeling, IL: Harlan Davidson, 1983), 39.

135. H. Paul Jeffers, *Commissioner Roosevelt: The Story of Theodore Roosevelt and the New York City Police, 1895–1897* (New York: John Wiley & Sons, 1994), 117.

136. Theodore Roosevelt, *An Autobiography* (New York: Charles Scribner's Sons, 1922), 175.

137. Jeffers, 117, 177.

138. Richards, 265.

139. Ibid., xiv.

140. George Lankevich, "The Grand Army of the Republic in New York State, 1865–1898" (PhD diss., Columbia University, 1967), xv, 9, 12, 15, 59.

141. McConnell, 15.

142. Lankevich, "The Grand Army," 24; "Public School Cadets," *Schools Devoted to the Public Schools and Educational Interests* 17 October 1895, 56.

143. "Public School Cadets," 56; "Military Instruction," *Schools Devoted to the Public Schools and Educational Interests* 26 December 1895, 141.

144. Egbert L. Viele, "The Moral Influence of Military Training" (unpublished manuscript, USMA Special Collections, 1883), 1.

145. Drew Gilpin Faust, *This Republic of Suffering: Death and the American Civil War* (New York: Vintage, 2009), xi–xiii.

146. Charles C. Calhoun, "The Political Culture: Public Life and the Conduct of Politics," in *The Gilded Age: Essays on the Origins of Modern America,* ed. Charles C. Calhoun (Wilmington, DE: Scholarly Resources, 1997), 189; Keller, 249–68.

147. Stuart McConnell, *Glorious Contentment: The Grand Army of the Republic, 1865–1900* (Chapel Hill: University of North Carolina Press, 1992), xiv, 126–65; Margaret Susan Thompson, *The "Spider Web": Congress and Lobbying in the Age of Grant* (Ithaca, NY: Cornell University Press, 1985), 259–63; Morton Keller, *Affairs of State: Public Life in Late Nineteenth Century America* (Cambridge, MA: Belknap Press of Harvard University Press, 1977), 35, 81–84, 207–208, 222–37.

148. Keller, 15, 83–84, 223, 235.

149. Rutherford B. Hayes rose to the rank of brevet major general after serving with the Ohio volunteers. James Garfield, also a volunteer from Ohio, made it to major general after the Battle

of Chickamauga. Chester Arthur was a lawyer and Republican from New York when Governor Edwin D. Morgan made him a brigadier general in charge of federal recruits mobilizing for the war. Benjamin Harrison became a brevet brigadier general in 1865 after answering the Indiana governor's call to lead the 70th Indiana Volunteers. Warner, 166–67, 221; Michael Beschloss, ed., *The American Heritage Illustrated History of the Presidents* (New York: Crown, 2000), 272, 294.

150. Keller, 262–63.

151. Heather Cox Richardson, *West from Appomattox: The Reconstruction of America after the Civil War* (New Haven: Yale University Press, 2007), 121–23.

152. James Bryce, *The American Commonwealth* (New York: Macmillan, 1888), 1:104–105.

153. Ibid., 105.

154. Keller, 259.

155. McConnell, 197, 198.

156. James Weinstein, "Organized Business and the City Commission and Manager Movements," *Journal of Southern History* 28.2 (May 1962): 166–68.

157. Eugene P. Moehring, *Public Works and the Patterns of Urban Real Estate Growth in Manhattan, 1835–1894* (New York: Arno, 1981), 375.

8. THE EMERGENCE OF MODERN AMERICA IN WEST POINT'S NEW YORK

1. USMA Association of Graduates, *Annual Reunion of the Association of Graduates of the United States Military Academy*, 1902, 3–5, 142. Viele died 22 April 1902. Ever the consummate member of New York society, he enjoyed a banquet at Delmonico's as his last meal. See "Death of Gen. E. L. Viele," *New York Times*, 23 April 1902, 9.

2. "Day of Glory for West Point," *New York Times*, 12 June 1902, 1.

3. Horace Porter, "Address by the Orator of the Day," in *The Centennial of the United States Military Academy at West Point, New York, 1802–1902* (Washington, DC: GPO, 1904), 1:31–32, 41. Horace Porter, class of 1860, was a Medal of Honor recipient and was the U.S. ambassador to France during the centennial celebration.

4. Kristin L. Hoganson, *Fighting for American Manhood: How Gender Politics Provoked the Spanish-American and Philippino-American Wars* (New Haven, CT: Yale University Press, 1998), 3–4, 101, 201.

5. Lance Betros, ed., *West Point: Two Centuries and Beyond* (Abilene, TX: McWhiney Foundation Press, 2004), 14.

6. Lance Betros, *Carved from Granite: West Point Since 1902* (College Station: Texas A&M University Press, 2012), xvi, 3.

7. Richard Hofstadter, *Anti-intellectualism in American Life* (New York: Alfred A. Knopf, 1963), 6, 195–98.

8. Arnold Bennett, *Your United States: Impressions of a First Visit* (New York: George H. Doran, 1912), 183–91; Richard Hofstadter, *The Age of Reform: From Bryan to F.D.R.* (New York: Alfred A. Knopf, 1955), 176–78.

9. George J. Lankevich, *New York City: A Short History* (New York: New York University Press, 2002), 122–28.

10. John S. D. Eisenhower, *Agent of Destiny: The Life and Times of General Winfield Scott* (New York: Free Press, 1997), 303.

11. Sean Wilentz, *Chants Democratic: New York City and the Rise of the American Working Class, 1788–1850* (New York: Oxford University Press, 1986), 117–18, 121–23.

12. Mark Aldrich, "Earnings of American Civil Engineers 1820–1859," *Journal of Economic History* 31.2 (June 1971): 412; *Minutes of the Croton Aqueduct Board of the City of New York, July 18, 1849 to April 9, 1870* (New York: Mail and Express Company, 1903), 172.

13. *Minutes of the Croton Aqueduct Board of the City of New York, July 18, 1849 to April 9, 1870*, 180.

14. For a contemporary study of income in the Victorian era, see Edward Young, *Labor in Europe and America: A Special Report of the Wages, the Cost of Subsistence and the Conditions of the Working Classes in Great Britain, France, Belgium, Germany, and Other Countries of Europe, and the United States and British America* (Philadelphia: S.A. George, 1875), 819–22.

15. James G. McGivern, *First Hundred Years of Engineering Education in the United States (1807–1907)* (Spokane, WA: Gonzaga University Press, 1960), 33.

16. American Society of Civil Engineers, *Transactions of the American Society of Civil Engineers* (New York: ASCE, 1886), 40:229, 234, 272–73, 330.

17. William B. Skelton, "West Point and Officer Professionalism, 1817–1877," in *West Point: Two Centuries and Beyond,* ed. Betros, 26.

18. Skelton, 31.

19. Burton J. Bledstein, *The Culture of Professionalism: The Middle Class and the Development of Higher Education in America* (New York: Norton, 1978), 85–86.

20. Edwin G. Burrows and Mike Wallace, *Gotham: A History of New York City to 1898* (New York: Oxford University Press, 1999), 1235.

21. David M. Scobey, *Empire City: The Making and Meaning of* the *New York City Landscape* (Philadelphia: Temple University Press, 2002), 265.

22. Burrows and Wallace, 1236; Bennett, 57; Scobey, 217–50; Michael Rawson, *Eden on the Charles: The Making of Boston* (Cambridge, MA: Harvard University Press, 2010), 277; William Cronon, *Nature's Metropolis: Chicago and the Great West* (New York: Norton, 1991), 305.

23. Sven Beckert, *The Monied Metropolis: New York City and the Consolidation of the American Bourgeoisie, 1850–1896* (New York: Cambridge University Press, 2001), 19–20.

24. Ibid., 333; Cronon, 43.

25. Bennett, 42–44.

26. Morton Keller, *Affairs of State: Public Life in Late Nineteenth Century America* (Cambridge, MA: Belknap Press of Harvard University Press, 1977), 2.

27. Ibid.

28. Howard P. Segal, *Technological Utopianism in American Culture* (Chicago: University of Chicago Press, 1985), 76.

29. Stephen R. Taffe, *Commanding the Army of the Potomac* (Lawrence: University Press of Kansas, 2006), 215–17.

30. Stuart McConnell, *Glorious Contentment: The Grand Army of the Republic, 1865–1900* (Chapel Hill: University of North Carolina Press, 1992), xiii, 25.

31. Brian C. Melton, *Sherman's Forgotten General: Henry W. Slocum* (Columbia: University of

Missouri Press, 2007), 244–45; New York City Department of Parks & Recreation website, accessed 3 August 2011 at http://www.nycgovparks.org/parks/grandarmyplaza/highlights/11844.

32. "Harlem Canal Parades," *New York Times,* 17 June 1895, 1; Chase Viele, "A Short Biography of Egbert L. Viele" (unpublished manuscript, USMA Special Collections, 1973), 53–58; "Resolution Offered at the Regular Encampment of Lafayette Post No. 140 in Observance of the Death of General Egbert L. Viele," 16 May 1902, USMA Special Collections.

33. Otto Eisenschiml, *The Celebrated Case of Fitz John Porter: An American Dreyfus Affair* (New York: Bobbs-Merrill, 1950), 316 (photo).

34. McConnell, 196, 201.

35. Kenneth T. Jackson, ed., *The Encyclopedia of New York City* (New Haven, CT: Yale University Press, 1995), 501.

36. Frederick Jackson Turner, *The Frontier in American History* (New York: Henry Holt, 1920), 33, 258.

37. Gunther Barth, *City People: The Rise of Modern City Culture in Nineteenth-Century America* (New York: Oxford University Press, 1980), 26; Raymond A. Mohl, *The New City: Urban America in the Industrial Age, 1860–1920* (Wheeling, IL: Harlan Davidson, 1985), 110–12.

38. Daniel Eli Burnstein, *Next to Godliness: Confronting Dirt and Despair in Progressive Era New York City* (Urbana: University of Illinois Press, 2006), 32–39.

39. Arthur S. Link and Richard L. McCormick, *Progressivism* (Wheeling, IL: Harlan Davidson, 1983), 23.

40. Hofstadter, *The Age of Reform,* 275.

41. Egbert L. Viele, *The Topography and Hydrology of New York* (New York: Robert Craighead, 1865), 1.

BIBLIOGRAPHY

PRIMARY SOURCES

United States Military Academy Special Collections, West Point, New York

William Wallace Burns Papers

George Washington Cullum Papers

Charles Champion Gilbert Papers

Ulysses S. Grant Papers

George S. Greene Papers

George Washington Whitfield Hazzard Papers

Andrew Atkinson Humphreys Papers

Gouverneur Kemble Papers

Dennis Hart Mahan Papers

George Brinton McClellan Papers

John Newton Papers

Henry Warner Slocum Papers

William Farrar Smith Papers

The West Point Thayer Papers, 1802–1872. Edited by Cindy Adams (Association of Graduates, 1965)

Egbert L. Viele Papers

Other Archive and Manuscript Collections

Craven Family Papers, New-York Historical Society, New York

Francis Vinton Greene Papers, MSS, New York Public Library, New York

George Sears Greene Collection, MSS, Rhode Island Historical Society

Alexander Hamilton, Jr., Letters, Papers 1823–1882, Misc. MSS, New-York Historical Society, New York

Alfred Thayer Mahan Papers, Library of Congress (microfiche)

Mayor's Papers, Municipal Archives and Record Center, New York

George Brinton McClellan Papers, Library of Congress (microfiche)

Fitz John Porter Papers, Library of Congress (microfiche)

Samuel J. Tilden Papers, MSS, New York Public Library, New York

West Side Association Minutes, New-York Historical Society, New York

Official Documents

American Society of Civil Engineers. *Proceedings of the American Society of Civil Engineers.* Vol. 1. Nov. 1873–Dec. 1875. New York: American Society of Civil Engineers, 1876.

———. *Proceedings of the American Society of Civil Engineers.* Vol. 6, Jan.–Dec. 1880. New York: ASCE, 1880.

———. *Transactions of the American Society of Civil Engineers* (Instituted 1852). Vol. 40. New York: ASCE, 1886.

———. *Transactions of the American Society of Civil Engineers.* Vol. 49. New York: ASCE, 1902.

Annual Report of the Secretary of War, 24 November 1828. *American State Papers, Military Affairs,* 4:2.

Biographical Directory of the United States Congress, 1774–Present. Accessed 7 February 2003 at http://bioguide.congress.gov/.

Board of Aldermen of the City of New York. *Annual Report of the Croton Aqueduct Department Made to the Common Council of the City of New York.* New York, 3 January 1855.

———. "Report of Mr. D. B. Douglass to the Commissioners for Supplying the City of New York with Pure and Wholesome Water," 1 February 1835. Document 44. Vol. 1 (1835), 403–33.

———. "Report of the Special Committee of the Board of Aldermen, Appointed to Investigate the Sufficiency of the Charges Alleged by His Honor the Mayor for the Removal of Messrs. Craven and Tappen." Document 18. Vol. 27, part 2. New York: Edmund Jones, 1860.

———. *Proceedings.* New York, 1849–71.

Board of Commissioners of the Central Park. *Documents of the Board of Commissioners of the Central Park for the Year Ending April 30, 1858.*

———. *Fourth Annual Report.* New York: Wm. C. Bryant, 1861.

———. *Minutes of the Proceedings of the Board of Commissioners of the Central Park.* New York, 1858–1869.

Board of Commissioners of the Department of Public Parks. *Annual Report.* New York, 1871–1874.

―――. *Minutes.* New York, 1871–1898.

Board of Visitors to the United States Military Academy. *Annual Report of the Board of Visitors to the United States Military Academy* (June 1824), 100.

―――. *Report of the Board of Visitors.* 24 June 1826.

―――. *Report of the Board of Visitors.* June 1867.

Citizens' Association of New York. *Report of the Council of Hygiene and Public Health of the Citizens' Association of New York upon the Sanitary Condition of the City.* 2nd ed. New York: D. Appleton, 1866.

Cullum, George W. *Biographical Register of the Officers and Graduates of the Military Academy at West Point, N.Y.* Vols. 1 and 2. 3rd ed. New York: Houghton Mifflin, 1891.

―――. *Biographical Register of the Officers and Graduates of the U.S. Military Academy at West Point.* Vol. 1. Cambridge, MA: Riverside Press, 1891.

Minutes of the Croton Aqueduct Board of the City of New York, July 18, 1849 to April 9, 1870. New York: Mail and Express Co., 1903.

New York State. *Documents of the Senate of the State of New York,* 1895.

U.S. Congress. *Congressional Record—House.* 49th Congress, 1st and 2nd Sessions. Washington, DC: Government Printing Office.

―――. *Debates and Proceedings in the Congress of the United States. 1834–1856.* Washington, DC: Gales and Seaton, 1856.

U.S. Army Corps of Engineers. *Analytical and Topical Index to the Reports of the Chief of Engineers and the Officers of the Corps of Engineers, United States Army, upon Works and Surveys for River and Harbor Improvement, 1866–1879.* Washington, DC: GPO, 1881.

―――. "Report of the Chief of Engineers," in *Report of the Secretary of War to the 41st Congress,* Vol. 2. Washington, DC: GPO, 1869.

United States Military Academy. *The Centennial of the United States Military Academy at West Point, New York.* 2 vols. Washington, DC: Government Printing Office, 1904.

―――. *Official Register of the Officers and Cadets of the United States Military Academy.* Multivolume. West Point, New York, 1818–1966.

―――. *Regulations of the U.S. Military Academy at West Point.* New York: J&J Harper, 1832.

―――. *Staff Records* 3:1842–45.

United States Military Academy Association of Graduates. *Annual Reunion of the Association of Graduates of the United States Military Academy at West Point, New York.* West Point, New York. 1870–1916.

―――. *Register of Graduates and Former Cadets of the United States Military Academy, Class of 1900 Centennial Edition.* West Point, New York, 2000.

U.S. War Department. *Report of the Chief of Engineers,* volume 2. Washington, DC: GPO, 1916.

Viele, Egbert L. *First Annual Report on the Improvement of the Central Park, New York, January 1, 1857.* Reprint, Washington, DC: McGrath Publishing.

Viele v. Gray. Supreme Court, New York County, New York, December Term 1859. In West Headnotes: 1859 WL 7727 (N.Y. Com.Pl.).

The War of Rebellion: A Compilation of the Official Records of the Union and Confederate Armies. 128 vols. Washington, DC: GPO, 1881–1901. Available online at Cornell University Making of America website, http://cdl.library.cornell.edu/cgi-bin /moa/.

Websites

Biographical Directory of the United States Congress, 1774–Present. http://bioguide.con gress.gov/.

NYC Department of Parks and Recreation. http://www.nycgovparks.org/.

U.S. Army Corps of Engineers, New York District. "The Conquest of Hell Gate." http://www.nan.usace.army.mil/whoweare/hellgate.pdf.

U.S. Census. http://factfinder.census.gov/home/saff/main.html?_lang=en.

Newspapers and Magazines

Army and Navy Journal

Assembly

Brooklyn Daily Eagle

Engineering News and American Contract Journal

Evening Mail

Evening Post

Harper's New Monthly Magazine and Harper's Magazine

Harper's Weekly

Mechanic's Magazine, Museum, Register, Journal, and Gazette

New York Daily News

New York Herald

New York Herald-Tribune

New York Morning Courier and New York Enquirer

New York Times

New York Tribune

Real Estate Record and Builders' Guide. Accessed via the Columbia University Libraries Digital Collections at http://ldpd.lamp.columbia.edu/rerecord/.

School: Devoted to the Public Schools and Educational Interests

Scientific American
Times Herald-Record

Books, Memoirs, Articles, and Correspondence

Beveridge, Charles E., Carolyn F. Hoffman and Kenneth Hawkins, eds. *The Papers of Frederick Law Olmsted.* Vol. 7. *Parks, Politics, and Patronage, 1874–1882.* Baltimore: Johns Hopkins University Press, 2007.

Beveridge, Charles E., and David Schuyler, eds. *The Papers of Frederick Law Olmsted.* Vol. 3: *Creating Central Park, 1857–1861.* Baltimore: Johns Hopkins University Press, 1983.

Bryant, William Cullen, II, and Thomas G. Voss, eds. *The Letters of William Cullen Bryant, 1809–1836.* Vol. 1. New York: Fordham University Press, 1975.

Craven, Alfred W. "President's Address to the Annual Convention of the American Society of Civil Engineers, 1870." In Wisely, *The American Civil Engineer, 1852–1974.*

Eastman, Seth. *Treatise on Topographical Drawing.* New York: Wiley and Putnam, 1837.

Grant, Ulysses S. *Personal Memoirs of U.S. Grant. Modern Library Edition.* Edited by Caleb Carr. Charles Webster and Company, 1885. Reprint, New York: Random House, 1999.

Greene, Francis Vinton. "Memoir of the Life and Services of George Sears Greene." New York, 1902.

Halleck, Henry. *Elements of Military Art and Science.* New York: D. Appleton, 1861.

Jervis, John B. *Description of the Croton Aqueduct.* New York: Slamm and Guion, 1842.

Johnson, Richard M. "Report on the Military Academy at West Point, by the Committee on Military Affairs, Submitted by Its Chairman, Col. R. M. Johnson," 17 May 1834, in *American Quarterly Review* 16 (Sept.–Dec. 1834): 358–75.

King, Charles. *A Memoir of the Construction, Cost, and Capacity of the Croton Aqueduct, Compiled from Official Documents.* New York: Charles King, 1843.

King, Rufus, to Christopher Gore, 22 June 1821. In *Life and Correspondence of Rufus King,* 6 vols. Edited by Charles R. King. New York: Putnam and Sons, 1900. 6:393–94.

Mahan, Alfred Thayer. *From Sail to Steam: Recollections of Naval Life.* New York: Harper & Brothers, 1907.

Mahan, Dennis Hart. *An Elementary Treatise on Advanced Guard, Out Post, and Detachment Service of Troops.* New York: John Wiley, 1863.

———. *A Treatise on Field Fortification, Containing Instructions on the Methods of Laying Out, Constructing, Defending, and Attacking Intrenchments, with the General Outlines Also of The Arrangement, The Attack and Defence of Permanent Fortifications.* New York: John Wiley, 1836; reprint, 1862.

Maury, Dabney Herndon. *Recollections of a Virginian in the Mexican, Indian, and Civil War.* New York: Charles Scribner's and Sons, 1894.

McAlpine, William J., and Egbert L. Viele. *The Opinion of Two Eminent Civil Engineers on Rapid Transit.* New York: Douglas Taylor, 1876.

Olmsted, Frederick Law Sr. *Forty Years of Landscape Architecture: Central Park.* Edited by Frederick Law Olmsted, Jr., and Theodora Kimball. Cambridge, MA: MIT Press, 1973.

Ranney, Victoria Post, ed. *The Papers of Frederick Law Olmsted.* Vol. 5. *The California Frontier, 1863–1865.* Baltimore: Johns Hopkins University Press, 1990.

Schofield, John McAllister. *Forty-Six Years in the Army.* 1897. Reprint, Norman: University of Oklahoma Press, 1998.

Sheridan, Philip H. *Personal Memoirs of P. H. Sheridan.* Vol. 1. New York: Charles L. Webster, 1888.

Smith, Gustavus W. *Life Insurance: The Practical Business.* New York: D. Van Nostrand, 1878.

Smith, William Farrar. *Autobiography of Major General William Farrar Smith, 1861–1864.* Edited by Herbert M. Schiller. Dayton, OH: Morningside, 1990.

Strong, George Templeton. *The Diary of George Templeton Strong.* 4 vols., Edited by Allan Nevins and Milton Halsey Thomas. New York: Macmillan, 1952.

Ticknor, George. *Life, Letters, and Journals of George Ticknor.* Vol. 1. Boston: James R. Osgood, 1877.

Turner, Charles W., ed. *The Education of Col. David Bullock Harris, C.S.A.: West Point Letters (1829–'35).* Verona, VA: McClure, 1984.

Viele, Egbert L. *The Arcade Under-Ground Railway.* New York, 1872.

———. "A Glimpse of Nature from My Veranda." *Harper's New Monthly Magazine* 57 (August 1878): 404–12.

———. *Handbook for Active Service; Containing Practical Instructions in Campaign Duties for the Use of Volunteers.* New York: D. Van Nostrand, 1861.

———. *Map of the City of New York, From the Battery to 80th Street, Showing the Original Topography of Manhattan Island.* Map to accompany testimony of Egbert Viele before the Sanitary Committee of the Senate. New York, 1865–1866.

———. *Map of the Lands Included in the Central Park from a Topographical Survey.* June 17th, 1856. New-York Historical Society.

———. "The Moral Influence of Military Training." Speech, Viele Papers, USMA Special Collections, ca. 1883.

———. "Navigation and Improvement of the Great Lakes and Basins of the North-West." Speech read at the Trans-Continental Convention, Oswego, NY, 6 October 1869.

———. *Plan for the Improvement of the Central Park.* Map. New York, 1856.

_____. "Prospect Park, Report, January 15, 1861," in *First Annual Report of the Commissioners of Prospect Park*, Brooklyn, NY, 28 January 1861.

_____. "Report on the Civic Cleanliness, and the Economical Disposition of the Refuse of Cities." New York: Edmund Jones, 1860.

———. *Robert A. Chesebrough's System of Locomotion for Elevated Railroads: Patented July 14, 1868.* New York: C. S. Westcott, 1869.

———. *Sanitary & Topographical Map of the City and Island of New York* ("Viele Water Map"). New York, 1865.

———. *The Topography and Hydrology of New York.* New York: Robert Craighead, 1865.

———. "Topography of New-York and Its Park System." In *The Memorial History of the City of New-York and the Hudson River Valley,* ed. James Grant Wilson. New York: New-York History, 1892, 4:551–60.

———. *The Transval of the City of New York.* New York: Johnson, 1880.

———. "A Trip with Lincoln, Chase, and Stanton." *Scribner's Monthly* 16.6 (October 1878): 813–23.

———. *The West End Plateau of the City of New York.* New York: Johnson and Pratt, 1879.

Viele, Teresa Griffin. *Following the Drum: A Glimpse of Frontier Life.* 1858. Reprint, Lincoln: University of Nebraska Press, 1984.

———. *Scrapbooks, 1870–1871.* 2 vols. Special Collections, Univ. of Delaware.

Wayland, Francis. *Report to the Corporation of Brown University, on Changes in the System of Collegiate Education.* Providence, RI: George H. Whitney, 1850.

Young, Edward. *Labor in Europe and America: A Special Report of the Wages, the Cost of Subsistence and the Conditions of the Working Classes in Great Britain, France, Belgium, Germany, and Other Countries of Europe, and the United States and British America.* Philadelphia: S.A. George, 1875.

SECONDARY SOURCES

Books and Articles

Abbott, Henry L. "Memoir of Dennis Hart Mahan: Read before the National Academy, November 7, 1878." *Memoirs of the National Academy of Sciences.* Vol. 2. Washington, DC: Government Printing Office, 1886.

Albion, Robert G. *The Rise of New York Port, 1815–1860.* New York: Charles Scribner's Sons, 1939.

Aldrich, Mark. "Earnings of American Civil Engineers, 1820–1859." *Journal of Economic History* 31.2 (June 1971): 407–19.

Allen, Oliver E. *The Tiger: The Rise and Fall of Tammany Hall.* New York: Addison-Wesley, 1993.

Ambinder, Tyler G. "Fernando Wood and New York City's Secession from the Union: A Political Appraisal." *New York History* 68 (January 1987): 67–92.

Ambrose, Stephen E. *Duty, Honor, Country: A History of West Point.* Baltimore: Johns Hopkins University Press, 1999.

————. *Nothing like It in the World: The Men Who Built the Transcontinental Railroad, 1863–1869.* New York: Touchstone, 2000.

Anders, Curt. *Injustice on Trial: Second Bull Run, General Fitz John Porter's Court-Martial and the Schofield Board Investigation That Restored His Good Name.* Zionsville, IN: Guild Press, 2002.

Anderson, Marvin J. "The Architectural Education of Nineteenth-Century American Engineers: Dennis Hart Mahan at West Point." *Journal of Society of Architectural Historians* 67.2 (2008): 222–47.

Angevine, Robert G. "Individuals, Organizations, and Engineering: U.S. Army Officers and the American Railroads, 1827–1838." *Technology and Culture* 42.2 (April 2001).

————. *The Railroad and the State: War, Politics, and Technology in Nineteenth-Century America.* Stanford, CA: Stanford University Press, 2004.

Angulo, A. J. "The Polytechnic Comes to America: How French Approaches to Science Instruction Influenced Mid-Nineteenth Century American Higher Education." *History of Science* 50, part 3, no. 168 (September 2012): 315–38.

Armstrong, Ellis L., Michael Robinson, and Suellen M. Hoy. *History of Public Works in the United States, 1776–1976.* Chicago: American Public Works Association, 1976.

Asbury, Herbert. *The Gangs of New York: An Informal History of the Underworld.* New York: Alfred A. Knopf, 1928.

Baker, Jean H. *Affairs of Party: The Political Culture of Northern Democrats in the Mid-Nineteenth Century.* New York: Fordham University Press, 1998.

Barth, Gunther. *City People: The Rise of Modern City Culture in Nineteenth-Century America.* New York: Oxford University Press, 1980.

Baumer, William Henry Jr. *Not All Warriors: Portraits of 19th Century West Pointers Who Gained Fame in Other than Military Fields.* 1941. Reprint, New York: Ayer, 1977.

Beckert, Sven. *The Monied Metropolis: New York City and the Consolidation of the American Bourgeoisie, 1850–1896.* New York: Cambridge University Press, 2001.

Bender, Thomas. *The Unfinished City: New York and the Metropolitan Idea.* New York: New Press, 2002.

Bennett, Arnold. *For Your United States: Impressions of a First Visit.* New York: George H. Doran, 1912.

Bernstein, Iver. *The New York City Draft Riots: Their Significance for American Society and Politics in the Age of the Civil War.* New York: Oxford University Press, 1990.

Bernstein, Peter L. *Wedding of the Waters: The Erie Canal and the Making of a Great Nation.* New York: Norton, 2005.

Betros, Lance, ed. *Carved from Granite: West Point Since 1902.* College Station: Texas A&M University Press, 2012.

———. *West Point: Two Centuries and Beyond.* Abilene, TX: McWhiney Foundation Press, 2004.

Blake, Nelson M. *Water for the Cities: A History of the Urban Water Supply Problem in the United States.* Syracuse, NY: Syracuse University Press, 1956.

Bledstein, Burton J. *The Culture of Professionalism: The Middle Class and the Development of Higher Education in America.* New York: Norton, 1978.

Bluestone, Daniel M. "From Promenade to Park: The Gregarious Origins of Brooklyn's Park Movement." *American Quarterly* 39.4 (Winter 1987): 529–50.

Blumin, Stuart M. *The Emergence of the Middle Class: Social Experience in the American City, 1760–1900.* New York: Cambridge University Press, 1989.

Bone, Kevin, Gina Pollara, and Albert F. Appleton. *Water-works: The Architecture and Engineering of the New York City Water Supply.* New York: Monacelli, 2006.

Bruce, Robert V. *The Launching of Modern American Science, 1846–1876.* Ithaca, NY: Cornell University Press, 1987.

Bryce, James. *The American Commonwealth.* Vol. 1. New York: Macmillan, 1888.

Burnstein, Daniel Eli. *Next to Godliness: Confronting Dirt and Despair in Progressive Era New York City.* Urbana: University of Illinois Press, 2006.

Burrows, Edwin G., and Mike Wallace. *Gotham: A History of New York City to 1898.* New York: Oxford University Press, 1999.

Buttenwieser, Ann L. *Manhattan Water-Bound: Manhattan's Waterfront from the Seventeenth Century to the Present.* 2nd ed. Syracuse, NY: Syracuse University Press, 1999.

Calhoun, Charles C., ed. *The Gilded Age: Essays on the Origins of Modern America.* Wilmington, DE: Scholarly Resources, 1997.

Calhoun, Daniel H. *The American Civil Engineer: Origins and Conflict.* Cambridge, MA: Harvard University Press, 1960.

Callow, Alexander. *The Tweed Ring.* New York: Oxford University Press, 1969.

Carman, Harry James, and Arthur W. Thompson. *A Guide to the Principal Sources for American Civilization, 1800–1900, in the City of New York: Manuscripts.* New York: Columbia University Press, 1960.

———. *A Guide to the Principal Sources for American Civilization, 1800–1900, in the City of New York: Printed Materials.* New York: Columbia University Press, 1962.

Carson, Hampton L. "Andrew Athinson Humphreys, Brigadier-General U.S. Army, Brevet Major-General, Chief of Engineers." *Proceedings of the American Philosophical Society* 22 (January 1885): 48–71.

Chandler, Alfred D. Jr. *The Visible Hand: The Managerial Revolution in American Business.* Cambridge, MA: The Belknap Press of Harvard University Press, 1977.

Cheape, Charles W. *Moving the Masses: Urban Public Transit in New York, Boston, and Philadelphia, 1880–1912.* Cambridge, MA: Harvard University Press, 1980.

Clark, Emmons. *History of the Seventh Regiment of New York, 1806–1889.* Vol. 2. New York, 1890.

Clausewitz, Carl von. *On War.* Edited and translated by Michael Howard and Peter Paret. Princeton, NJ: Princeton University Press, 1984.

Cohen, Paul E., and Robert T. Augustyn. *Manhattan in Maps, 1527–1995.* New York: Rizzoli International, 1997.

Cornog, Evan. "Whig Party." In *The Encyclopedia of New York City,* Edited by Kenneth T. Jackson. New Haven, CT: Yale University Press, 1995, 1257.

Costello, Albert E. *Our Police Protectors: History of the New York Police from the Earliest Period to the Present Time.* New York: Chas. F. Roper, 1885.

Cox, John. *Culp's Hill: The Attack and Defense of the Union Flank, July 2, 1863 (Battleground America).* Cambridge, MA: Da Capo, 2003.

Crackel, Theodore J. *West Point: A Bicentennial History.* Lawrence: University Press of Kansas, 2002.

Croly, Herbert. "New York as the American Metropolis." *Architectural Record* 13 (1903): 193–206.

Cronon, William. *Nature's Metropolis: Chicago and the Great West.* New York: Norton, 1991.

Cunliffe, Marcus. *Soldiers and Civilians: The Martial Spirit in America, 1775–1865.* London: Erye & Spottiswoode, 1968.

Curry, Leonard P. *The Corporate City: The American City as a Political Entity, 1800–1850.* Westport, CT: Greenwood, 1997.

Dilts, James D. *The Great Road: The Building of the Baltimore and Ohio, The Nation's First Railroad, 1828–1853.* Stanford, CA: Stanford University Press, 1993.

Domosh, Mona. *Invented Cities: The Creation of Landscape in Nineteenth-Century New York and Boston.* New Haven, CT: Yale University Press, 1996.

Dupuy, R. Ernest. *Men of West Point: The First 150 Years of the United States Military Academy.* New York: William Sloane Associates, 1951.

———. *Where They Have Trod: The West Point Tradition in American Life.* New York: Frederick A. Stokes, 1940.

Eisenhower, John S. D. *Agent of Destiny: The Life and Times of General Winfield Scott.* New York: Free Press, 1997.

Eisenschiml, Otto. *The Celebrated Case of Fitz John Porter: An American Dreyfus Affair.* New York: Bobbs-Merrill, 1950.

Endler, James R. *Other Leaders, Other Heroes: West Point's Legacy to America beyond the Field of Battle.* Westport, CT: Praeger, 1998.

Faust, Drew Gilpin. *This Republic of Suffering: Death and the American Civil War*. New York: Vintage, 2009.

Federal Writers' Project. *New York City Guide*. New York: Random House, 1939.

Fein, Albert. *Landscape into Cityscape: Frederick Law Olmsted's Plans for a Greater New York City*. Ithaca, NY: Cornell University Press, 1968.

Forster, John. *The Life of Charles Dickens*. Vol. 3, *1852–1879*. Philadelphia: J. B. Lippincott, 1874.

Gandy, Matthew. "The Bacteriological City and Its Discontents." *Historical Geography* 34 (2006): 14–25.

Gbondo, Francis. "Engineering History: Julius Walker Adams (1812–1899)." *American Society of Civil Engineers San Bernardino & Riverside Counties Branch Newsletter*. November 2009, 2.

Geissler, Suzanne. "Professor Dennis Mahan Speaks Out on West Point Chapel Issues, 1850." *Journal of Military History* 69.2 (April 2005): 505–19.

Glaeser, Edward L. "Urban Colossus: Why New York Is America's Largest City." *Federal Reserve Bank of New York Economic Policy Review* 11.2 (2005): 7–24.

Goldman, Joanne Abel. *Building New York's Sewers: Developing Mechanisms of Urban Management*. West Lafayette, IN: Purdue University Press, 1997.

Green, Jennifer R. *Military Education and the Emerging Middle Class in the Old South*. New York: Cambridge University Press, 2008.

Griffith, Ernest Stacy. *A History of American City Government: The Conspicuous Failure, 1870–1900*. Washington, DC: University Press of America, 1983.

Griffith, Ernest Stacey, and Charles R. Adrian. *A History of American City Government: The Formation Traditions, 1775–1870*. Washington, DC: University Press of America, 1983.

Gronowicz, Anthony. *Race and Class Politics in New York City before the Civil War*. Boston: Northeastern University Press, 1998.

Hacker, Barton C. *American Military Technology: The Life Story of a Technology*. Westport, CT: Greenwood, 2006.

———. "Engineering a New Order: Military Institutions, Technical Education, and the Rise of the Industrial State." *Technology and Culture* 34.1 (January 1993): 1–27.

Hall, Henry ed. *America's Successful Men of Affairs: An Encyclopedia of Contemporaneous Biography, volume 1*. New York: The New York Tribune, 1895.

Hammack, David C. *Power and Society: Greater New York at the Turn of the Century*. New York: Russell Sage Foundation, 1982.

Hesseltine, William Best, and Hazel C. Wolf. *The Blue and the Gray on the Nile*. Chicago: University of Chicago Press, 1961.

Hill, Forest G. *Roads, Rails, & Waterways: The Army Engineers and Early Transportation*. Norman: University of Oklahoma Press, 1957.

Hochfelder, David. *The Telegraph in America, 1832–1920*. Baltimore: Johns Hopkins University Press, 2012.

Hofstadter, Richard. *The Age of Reform: From Bryan to F.D.R.* New York: Alfred A. Knopf, 1955.

———. *Anti-intellectualism in American Life*. New York: Alfred A. Knopf, 1963.

Hoganson, Kristin L. *Consumers' Imperium: The Global Production of American Domesticity, 1865–1920*. Chapel Hill: University of North Carolina Press, 2007.

———. *Fighting for American Manhood: How Gender Politics Provoked the Spanish-American and Philippino-American Wars*. New Haven, CT: Yale University Press, 1998.

Hope, Ian C. *A Scientific Way of War: Antebellum Military Science, West Point, and the Origins of American Military Thought*. Lincoln: University of Nebraska Press, 2015.

Howe, Daniel Walker. *What Hath God Wrought: The Transformation of America, 1815–1848*. New York: Oxford University Press, 2007.

Hoy, Suellen. *Chasing Dirt: The American Pursuit of Cleanliness*. New York: Oxford University Press, 1995.

Hsieh, Wayne Wei-siang. *West Pointers and the Civil War: The Old Army in War and Peace*. Chapel Hill: University of North Carolina Press, 2009.

Hudson, Leonne M. *The Odyssey of a Southerner: The Life and Times of Gustavus Woodson Smith*. Macon, GA: Mercer University Press, 1998.

Hunt, Charles Warren. "Address at the Annual Convention at Washington, D.C., May 20th, 1902: The First Fifty Years of the American Society of Civil Engineers: 1852–1902." *Transactions of the American Society of Civil Engineers* 48. New York, 1902.

———. *Historical Sketch of the American Society of Civil Engineers*. New York, 1897.

Hunter, Robert F., and Edwin L. Dooley. *Claudius Crozet: French Engineer in America, 1790–1864*. Charlottesville: University Press of Virginia, 1989.

Huntington, Samuel P. *The Soldier and the State: The Theory and Politics of Civil-Military Relations*. Cambridge, MA: The Belknap Press of Harvard University Press, 1985.

Jackson, Kenneth T. *Crabgrass Frontier: The Suburbanization of the United States*. New York: Oxford University Press, 1985.

———, ed. *The Encyclopedia of New York City*. New Haven, CT: Yale University Press, 1995.

Jackson, Kenneth T., and David S. Dunbar, eds. *Empire City: New York through the Centuries*. New York: Columbia University Press, 2002.

Jeffers, H. Paul. *Commissioner Roosevelt: The Story of Theodore Roosevelt and the New York City Police, 1895–1897*. New York: John Wiley & Sons, 1994.

Jordan, David M. *"Happiness Is Not My Companion: The Life of General G.K. Warren*. Indianapolis: Indiana University Press, 2001.

Jordan, Ervin L. *Black Confederates and Afro-Yankees in Civil War Virginia*. Charlottesville: University Press of Virginia, 1995.

Katz, Irving. *August Belmont: A Political Biography*. New York: Columbia University Press, 1968.

Keller, Morton. *Affairs of State: Public Life in Late Nineteenth Century America*. Cambridge, MA: The Belknap Press of Harvard University Press, 1977.

Kemp, Emory L., and Edward Winant. "John Jervis and the Hydraulic Design of the Old Croton Aqueduct." *Canal History & Technology Proceedings* 22 (2003): 56–78.

Kershner, James William. *Sylvanus Thayer: A Biography*. New York: Arno, 1982.

Kessner, Thomas. *Capital City: New York City and the Men behind America's Rise to Economic Dominance, 1860–1900*. New York: Simon and Schuster, 2003.

Klawonn, Marion J. *Cradle of the Corps: A History of the New York District U.S. Army Corps of Engineers, 1775–1975*. New York: U.S. Army Engineer Corps, 1977.

Kobrick, Jason. "No Army Inspired: The Failure of Nationalism at Antebellum West Point." *Concept: An Interdisciplinary Journal of Graduate Studies*. Villanova University (2004): 1–18.

Koeppel, Gerard T. *Water for Gotham: A History*. Princeton, NJ: Princeton University Press, 2000.

Kornblum, William. *At Sea in the City: New York from the Water's Edge*. Chapel Hill, NC: Algonquin, 2002.

Kowsky, Francis R. *Country, Park, and City: The Architecture and Life of Calvert Vaux*. New York: Oxford University Press, 1998.

Kuhn, Reinhard Clifford. *The Return to Reality: A Study of Francis Viele-Griffin*. Paris: Librairie Minard, 1962.

Lankevich, George J. *New York City: A Short History*. New York: New York University Press, 2002.

Lankton, Larry D. *The "Practicable" Engineer: John B. Jervis and the Old Croton Aqueduct*. Chicago: Public Works Historical Society, 1977.

Lathrop, George Parsons, and Rose Hawthorne Lathrop. *Story of Courage: Annals of the Georgetown Convent of the Visitation of the Blessed Virgin Mary*. Cambridge, MA: Riverside, 1894.

Layton, Edwin T. *The Revolt of the Engineers: Social Responsibility and the American Engineering Profession*. Cleveland: Case Western Reserve University Press, 1971.

Lears, Jackson. *No Place of Grace: Antimodernism and the Transformation of American Culture, 1880–1920*. New York: Pantheon, 1983.

———. *Rebirth of a Nation: The Making of Modern America, 1877–1920*. New York: Harper Perennial, 2010.

Leeman, William P. *The Long Road to Annapolis: The Founding of the Naval Academy and the Emerging American Republic*. Chapel Hill: University of North Carolina Press, 2010.

Link, Arthur S., and Richard L. McCormick. *Progressivism*. Wheeling, IL: Harlan Davidson, 1983.

Linn, Brian M. *The Echo of Battle: The Army's Way of War.* Cambridge, MA: Harvard University Press, 2007.

Logel, Jon Scott. "Party, Park, and a Pyramid: Egbert L. Viele and the Creation of Central Park." *De Halve Maen* 75.4 (Winter 2002): 63–68.

Marcus, Alan I., and Howard P. Segal. *Technology in America: A Brief History.* New York: Harcourt Brace Jovanovich, 1989.

Marszalek, John F. *Commander of All Lincoln's Armies: A Life of General Henry W. Halleck.* Cambridge, MA: The Belknap Press of Harvard University Press, 2004.

Martin, Justin. *Genius of Place: The Life of Frederick Law Olmsted.* Cambridge, MA: Da Capo, 2011.

McConnell, Stuart. *Glorious Contentment: The Grand Army of the Republic, 1865–1900.* Chapel Hill: University of North Carolina Press, 1992.

McCullough, David. *The Great Bridge: The Epic Story of the Building of the Brooklyn Bridge.* New York: Simon and Schuster, 1982.

———. *The Greater Journey: Americans in Paris.* New York: Simon and Schuster, 2011.

McDonald, Robert M. S., ed. *Thomas Jefferson's Academy: Founding West Point.* Charlottesville: University of Virginia Press, 2004.

McGivern, James G, *First Hundred Years of Engineering Education in the United States (1807–1907).* Spokane, WA: Gonzaga University Press, 1960.

McIntire, Samuel B. "Echoes of the Past." *Army and Navy Journal* 39 (14 June 1902): 1026.

McKay, Ernest A. *The Civil War and New York City.* Syracuse, NY: Syracuse University Press, 1990.

McMaster, R. K. *West Point's Contribution to Education, 1802–1952.* New York: McMath, 1952.

McNeur, Catherine. *Taming Manhattan: Environmental Battles in the Antebellum City.* Cambridge, MA: Harvard University Press, 2014.

McPherson, James M. *Battle Cry of Freedom: The Civil War Era.* New York: Oxford University Press, 1988.

———. *Tried by War: Abraham Lincoln as Commander in Chief.* New York: Penguin. 2008.

Meier, Hugo A. "Technology and Democracy, 1800–1860." *Mississippi Valley Historical Review* 44.2 (1957): 618–40.

Melosi, Martin V. *The Sanitary City: Environmental Services in Urban America from Colonial Times to the Present.* Pittsburgh: University of Pittsburgh Press, 2008. Abridged edition of *The Sanitary City: Urban Infrastructure in America from Colonial Times to the Present.* Baltimore: Johns Hopkins University Press, 2000.

Melton, Brian C. *Sherman's Forgotten General: Henry W. Slocum.* Columbia: University of Missouri Press, 2007.

Merritt, Raymond H. *Engineering in American Society, 1850–1875.* Lexington: University Press of Kentucky, 1969.

Moehring, Eugene P. *Public Works and the Patterns of Urban Real Estate Growth in Manhattan, 1835–1894.* New York: Arno, 1981.

Mohl, Raymond A. *The New City: Urban America in the Industrial Age, 1860–1920.* Wheeling, IL: Harlan Davidson, 1985.

Morrison, James L. Jr. *"The Best School": West Point, 1833–1866.* Kent, OH: Kent State University Press, 1998.

———. "The Struggle between Sectionalism and Nationalism." *Civil War History* 19 (June 1973): 143.

Moten, Matthew. *The Delafield Commission and the American Military Profession.* College Station: Texas A&M University Press, 2000.

Mountcastle, Clay. *Punitive War: Confederate Guerrillas and Union Reprisals.* Lawrence: University Press of Kansas, 2009.

Mowbray, Jay Henry, ed. *Representative Men of New York: A Record of Their Achievements.* Vol. 2. New York: New York Press, 1898.

Mushkat, Jerome. *Fernando Wood: A Political Biography.* Kent, OH: Kent State University Press, 1990.

The National Cyclopaedia of American Biography. Vol. 2. New York: James T. White, 1921.

The National Cyclopedia of American Biography. Vol. 9. 1907.

Nevins, Allan. *The Evening Post: A Century of Journalism.* New York: Boni and Liveright, 1922.

Noble, David F. *The Religion of Technology: The Divinity of Man and the Spirit of Invention.* New York: Alfred A. Knopf, 1997.

Norris, L. David, James C. Milligan, and Odie B. Faulk. *William H. Emory: Soldier Scientist.* Tucson: University of Arizona Press, 1998.

Ogle, Maureen. "Water Supply, Waste Disposal, and the Culture of Privatism in the Mid-Nineteenth-Century American City." *Journal of Urban History* 25 (March 1999): 321–47.

Palmer, David W. *The Forgotten Hero of Gettysburg: A Biography of General George Sears Greene.* Philadelphia: Xlibris, 2005.

Pappas, George S. *To the Point: The United States Military Academy, 1802–1902.* Westport, CT: Praeger, 1993.

Parkman, Aubrey. *Army Engineers in New England: The Military and Civil Work of the Corps of Engineers in New England, 1775–1975.* Waltham, MA: U.S. Army Corps of Engineers, 1978.

Peterson, Jon A. "The Impact of Sanitary Reform upon American Urban Planning, 1840–1890." *Journal of Social History* 13.1 (Autumn 1979): 83–103.

Porter, Glenn. *The Rise of Big Business, 1860–1920.* 2nd ed. Wheeling, IL: Harlan Davidson, 1992.

Pryor, Elizabeth Brown. *Reading the Man: A Portrait of Robert E. Lee through His Private Letters.* New York: Penguin, 2008.

Quigley, David. *Second Founding: New York City, Reconstruction, and the Making of American Democracy*. New York: Hill and Wang, 2004.

Rae, John, and Rudi Volti. *The Engineer in History*. Revised ed. New York: Peter Lang, 2001.

Rawson, Michael. *Eden on the Charles: The Making of Boston*. Cambridge, MA: Harvard University Press, 2010.

Reed, Henry Hope, and Sophia Duckworth. *Central Park: A History and a Guide*. New York: Clarkson N. Potter, 1967.

Reel, David M. "The Drawing Curriculum at the U.S. Military Academy during the 19th Century." In *West Point Points West*. Denver: Institute of Western American Art, 2002, 51–60.

Reier, Sharon. *The Bridges of New York*. New York: Quadrant, 2000.

Reuss, Martin. "Andrew A. Humphreys and the Development of Hydraulic Engineering: Politics and Technology in the Army Corps of Engineers." *Technology and Culture* 26.1 (January 1985): 1–33.

Revell, Keith D. *Building Gotham: Civic Culture and Public Policy in New York City, 1898–1938*. Baltimore: Johns Hopkins University Press, 2005.

Reynolds, Terry S. "The Education of Engineers in America before the Morrill Act of 1862." *History of Education Quarterly* 32.4 (Winter 1992): 459–82.

Richardson, Heather Cox. *West from Appomattox: The Reconstruction of America after the Civil War*. New Haven, CT: Yale University Press, 2007.

Richardson, James F. *The New York Police: Colonial Times to 1901*. New York: Oxford University Press, 1970.

Rider, Fremont. *Rider's New York City*. New York, 1916.

Roosevelt, Theodore. *An Autobiography*. Charles Scribner's Sons, 1922.

———. *New York*. New York: Longmans, Green, 1891.

Rosenberg, Charles E. *The Cholera Years: The United States in 1832, 1849, and 1866*. Chicago: University of Chicago Press, 1987.

Rosenzweig, Roy, and Elizabeth Blackmar. *The Park and the People: A History of Central Park*. Ithaca, NY: Cornell University Press, 1992.

Rybczynski, Witold. *A Clearing in the Distance: Frederick Law Olmsted and America in the Nineteenth Century*. New York: Scribner, 1999.

Salwen, Peter. "Past Tense: Soldier of Misfortune." *New York Alive* (Jan.–Feb. 1990): 14–16.

———. *Upper West Side Story: A History and Guide*. New York: Abbeville, 1989.

Schaffer, Daniel, ed. *Two Centuries of American Planning*. Baltimore: Johns Hopkins University Press, 1988.

Schultz, Stanley K., and Clay McShane. "To Engineer the Metropolis: Sewers, Sanitation, and City Planning in Late-Nineteenth-Century America." *Journal of American History* 65.2 (September 1978): 389–411.

Schwager, E. "From Petroleum Jelly to Riches." *Drug News and Perspectives* 11.2 (1998): 127.

Scobey, David M. *Empire City: The Making and Meaning of the New York City Landscape.* Philadelphia: Temple University Press, 2002.

Sears, Stephen W. *George B. McClellan: The Young Napoleon.* New York: Ticknor and Fields, 1988.

Segal, Howard P. *Technological Utopianism in American Culture.* Chicago: University of Chicago Press, 1985.

Seidule, J. T. "'Treason Is Treason': Civil War Memory at West Point, 1861–1902." *Journal of Military History* 76.2 (2012): 427–545.

Sellers, Charles. *The Market Revolution: Jacksonian America, 1815–1846.* New York: Oxford University Press, 1991.

Shackleton, Robert. *The Book of New York.* Philadelphia: Penn Publishing, 1917.

Shallat, Todd A. *Structures in the Stream: Water, Science, and the Rise of the U.S. Army Corps of Engineers.* Austin: University of Texas Press, 1994.

Skelton, William B. *An American Profession of Arms: The Army Officer Corps, 1784–1861.* Lawrence: University Press of Kansas, 1992.

———. "Samuel P. Huntington and the Roots of the American Military Tradition." *Journal of Military History* 60.2 (April 1996): 325–38.

Slocum, Charles Elihu. *The Life and Services of Major General Henry Warner Slocum.* Toledo, OH: Slocum, 1913.

Smiles, Samuel. *The Life of Thomas Telford, Civil Engineer: With an Introductory History of Roads and Travelling in Great Britain.* London: John Murray, Albemarle Street, 1867.

Smith, Carl S. *City Water, City Life: Water and the Infrastructure of Ideas in Urbanizing Philadelphia, Boston, and Chicago.* Chicago: University of Chicago Press, 2013.

Smith, Jean Edward. *Grant.* New York: Simon and Schuster, 2001.

Soll, David. *Empire of Water: An Environmental and Political History of the New York City Water Supply.* Ithaca, NY: Cornell University Press, 2013.

Spann, Edward K. *Gotham at War: New York City, 1860–1865.* Wilmington, DE: Scholarly Resources, 2002.

———. *The New Metropolis: New York City, 1840–1857.* New York: Columbia University Press, 1981.

Steffens, Lincoln. "New York: Good Government in Danger." *McClure's Magazine* 22.1 (November 1903): 84–92.

Sullivan, Louis H. *The Autobiography of an Idea.* New York: American Institute of Architects, 1924.

Taffe, Stephen R. *Commanding the Army of the Potomac.* Lawrence: University Press of Kansas, 2006.

Teaford, Jon C. *The Unheralded Triumph: City Government in America, 1870–1900.* Baltimore: Johns Hopkins University Press, 1984.

Thompson, E. P. *The Making of the English Working Class.* New York: Vintage, 1966.

Thompson, Margaret Susan. *The "Spider Web": Congress and Lobbying in the Age of Grant.* Ithaca, NY: Cornell University Press, 1985.

Tillman, Samuel E. "The Academic History of the Military Academy, 1802–1902." In *The Centennial of the United States Military Academy at West Point, New York, 1802–1902.* Washington, DC: GPO, 1904, 1:353.

Trachtenberg, Alan. *The Incorporation of America: Culture and Society in the Gilded Age.* New York: Hill and Wang, 1982.

Tuckerman, Elise Strother. *The Pendulum Swings.* New York: Vantage, 1962.

Turner, Frederick Jackson. *The Frontier in American History.* New York: Henry Holt, 1920.

Upton, Dell. *Another City: Urban Life and* Urban *Spaces in the New American Republic.* New Haven, CT: Yale University Press, 2008.

Viele, Chase. "America's Pyramid on-the-Hudson." *Assembly* (December 1973): 20–23, 35.

———. "The Knickerbockers of Upstate New York." *De Halve Maen* 47 (October 1972): 1–2.

———. "A Short Biography of Egbert L. Viele." West Point, NY: United States Military Academy Special Collections, 1973. Unpublished.

Viele, Herman Knickerbocker. "A Sketch of the Life of General Egbert L. Viele" in *The New York Genealogical and Biological Record.* New York: New York Genealogical and Biographical Society. 34.1 (January 1903).

Viele, Kathlyne Knickerbocker. *Viele, 1659–1909: Two Hundred and Fifty Years with a Dutch Family of New York.* New York: Tobias A. Wright, 1909.

Vitiello, Domenic. "Monopolizing the Metropolis: Gilded Age Growth Machines and Power in American Urbanization." *Planning Perspectives* 28.1 (January 2013): 71–90.

Ward, William Hayes. *Abraham Lincoln: Tributes from His Associates, Reminiscences of Soldiers, Statesmen and Citizens.* New York: Thomas Y. Crowell, 1895.

Warner, Ezra J. *Generals in Blue: Lives of the Union Commanders.* Baton Rouge: Louisiana State University Press, 1964.

Watson, Samuel J. "Historiographical Essay: Continuity in Civil-Military Relations and Expertise: The U.S. Army before the Civil War." *Journal of Military History* 75.1 (January 2011): 221–50.

———. "How the Army Became Accepted: West Point Socialization, Military Accountability, and the Nation-State during the Jacksonian Era." *American Nineteenth Century History* 7 (June 2006): 219–51.

Wegmann, Edward. *The Water-Supply of the City of New York, 1658–1895.* New York: John Wiley & Sons, 1896.

Weidner, Charles H. *Water for a City: A History of New York City's Problem from the Beginning to the Delaware River System.* New Brunswick, NJ: Rutgers University Press, 1974.

Weigley, Russell F. "American Strategy from Its Beginnings through the First World War." In *Makers of Modern Strategy: From Machiavelli to the Nuclear Age*. Princeton, NJ: Princeton University Press, 1986, 408–43.

Weigold, Marilyn E. *Silent Builder: Emily Warren Roebling and the Brooklyn Bridge*. New York: Associated Faculty Press, 1984.

Weinstein, James. "Organized Business and the City Commission and Manager Movements." *Journal of Southern History* 28.2 (May 1962): 166–82.

White, Richard. *Railroaded: The Transcontinentals and the Making of Modern America*. New York: Norton, 2011.

Wiebe, Robert H. *The Search for Order, 1877–1920*. New York: Hill and Wang, 1967.

Wilentz, Sean. *Chants Democratic: New York City and the Rise of the American Working Class, 1788–1850*. New York: Oxford University Press, 1986.

Williams, T. Harry. *The History of American Wars from Colonial Times to World War I*. New York: Alfred A. Knopf, 1981.

Wilson, James Harrison. *Life and Service of William Farrar Smith*. Wilmington, DE: John M. Rogers, 1904.

Wisely, William H. *The American Civil Engineer, 1852–1974: The History, Traditions, and Development of the American Society of Civil Engineers*. New York: American Society of Civil Engineers, 1974.

Film

Burns, Ken. *Brooklyn Bridge*. PBS documentary film. New York: Florentine Films and WETA, 1982.

Dissertations

Bonura, Michael Andrew. "French Thought and the American Military Mind: A History of French Influence on the American Way of Warfare from 1814 through 1941." PhD diss., Florida State University, 2008.

Griess, Thomas E. "Dennis Hart Mahan: West Point Professor and Advocate of Military Professionalism, 1830–1871." PhD diss., Duke University, 1968.

Lankevich, George. "The Grand Army of the Republic in New York State, 1865–1898." PhD diss., Columbia University, 1967.

Manning, Larry. "The Contribution of Sylvanus Thayer and the United States Military Academy to Engineering Programs in the United States." PhD diss., Texas A&M University, 2003.

Molloy, Peter Michael. "Technical Education and the Young Republic: West Point as America's Ecole Polytechnique, 1802–1833." PhD diss., Brown University, 1975.

Nienkamp, Paul Keith. "A Culture of Technical Knowledge: Professionalizing Science and Engineering Education in Late-nineteenth Century America." PhD diss., Iowa State University, 2008.

INDEX

Italics indicate pages with illustrations.